ICE DESTRUCTION

GLACIOLOGY AND QUATERNARY GEOLOGY

A Series of Books

ICE DESTRUCTION
Methods and Technology

by

V. V. BOGORODSKY, V. P. GAVRILO,

and

O. A. NEDOSHIVIN

*Arctic and Antarctic Scientific Research
Institute, Leningrad, U.S.S.R.*

Translated from the Russian by

M. B. ROSENBERG

D. REIDEL PUBLISHING COMPANY

A MEMBER OF THE KLUWER  ACADEMIC PUBLISHERS GROUP

DORDRECHT / BOSTON / LANCASTER / TOKYO

Library of Congress Cataloging-in-Publication Data

L.C. Data appear on a separate card.

L.C. No. 86-31509.

ISBN 90-277-2229-3

Published by D. Reidel Publishing Company,
P.O. Box 17, 3300 AA Dordrecht, Holland.

Sold and distributed in the U.S.A. and Canada
by Kluwer Academic Publishers,
101 Philip Drive, Norwell, MA 02061, U.S.A.

In all other countries, sold and distributed
by Kluwer Academic Publishers Group,
P.O. Box 322, 3300 AH Dordrecht, Holland.

Originally published in 1983 by Gidrometeoizdat under the title
Raztushenie L'da - Metodi, Technicheskie, Sredstva

All Rights Reserved
© 1987 by D. Reidel Publishing Company, Dordrecht, Holland
No part of the material protected by this copyright notice may be reproduced or
utilized in any form or by any means, electronic or mechanical,
including photocopying, recording or by any information storage and
retrieval system, without written permission from the copyright owner

Printed in The Netherlands

CONTENTS

PREFACE	vii
1. MECHANICAL DESTRUCTION OF ICE	1
1.1. Perforating and Breaking	3
1.2. Cutting, Milling and Chipping	23
1.3. Drilling	51
1.4. Hydraulic Jet Methods	89
1.5. Explosive Methods	92
1.6. Inventions for Mechanical Destruction of Ice	108
2. THERMAL DESTRUCTION OF ICE	123
2.1. Blackening	124
2.2. Pneumatic and Hydrodynamic Methods	127
2.3. Steam-Water-Air Methods	147
2.4. Gas-Thermal Methods	153
2.5. Electrothermal Methods	158
2.6. Inventions for Thermal Destruction of Ice	165
3. CHEMICAL DESTRUCTION OF ICE	177
3.1. Inventions for Chemical Destruction of Ice	182
4. ELECTROPHYSICAL DESTRUCTION OF ICE	184
4.1. Electric Pulse Technique	184
4.2. Electric Hydraulic Technique	185
4.3. Laser Technique	186
4.4. Electromagnetic Field, Radioactive Radiation, Supersonic Waves	187
4.5. Inventions for Electrophysical Destruction of Ice	190
5. COMBINED DESTRUCTION OF ICE	191
5.1. Inventions for Combined Destruction of Ice	195
REFERENCES	198
SUBJECT INDEX	213

PREFACE

The problem of ice destruction comes most frequently to our attention in engineering glaciology and ice engineering because it is essential in the solution of many problems in the polar regions of the Earth.

Ice destruction (like the destruction of any other material, in principle) is a complex problem at the junction of solid-state physics, continuum mechanics, and materials science. Ice, particularly sea ice, is characterized by known anomalies that can be explained by the simultaneous occurrence of solid, liquid and gaseous phases. Even minor temperature fluctuations cause changes in the relationship of these phases and, as a consequence, change the physico-mechanical properties of ice. New hydraulic engineering tasks, associated with the destruction of such a complex material, demand continuous improvement of methods and techniques. The present authors have brought these together in a form which is convenient for a wide range of users. This book covers only local ice destruction, by means other than icebreakers, requiring comparatively low consumption of power in proportion to the volume and mass of destroyed ice. Problems of natural ice destruction under the influence of solar radiation, tidal, wind and wave factors are not discussed. Mechanical and thermal methods were the first of many to be used for ice destruction. Their application has involved a greater number of techniques, so the first two chapters are the longest. Lists of inventions for ice destruction patented in various countries are given at the end of each chapter. This approach is very valuable because it reveals new prospects in solving practical problems. V.V. Bogorodsky (a corresponding member of the USSR Academy of Sciences), V.P. Gavrilo (a candidate of physico-mathematical sciences, and O.A. Nedoshivin (a candidate of geographical sciences), have discussed complex problems of ice destruction from a modern physico-technical viewpoint.

This book on ice destruction will be of interest to a wide circle of readers and will contribute to further progress in engineering glaciology and ice engineering.

Chapter 1

MECHANICAL DESTRUCTION OF ICE

The destruction of any solid body is associated primarily with its strength, i.e. the ability of the material to endure various mechanical loads and non-uniform effects of physical fields under certain conditions and limits without destruction. The strengh of ice, in particular, depends greatly on the variety of its structural specifics, and is affected greatly by external conditions: the character of loads, thermal regime, aggressiveness of the environment, superficial effects, etc. Natural ice comprises numerous defects, ranging from microscopic and submicroscopic defects to large pores and trunk cracks.

The particular feature of ice, as compared to other materials, is that ice under natural conditions exists at a temperature close to the melting point. It will not react chemically with admixtures, and is characterized by a comparatively large crystalline structure. Research demonstrates that ice has a low limit of elasticity and manifests clearly expressed rheological properties even if these cause a reduction of strength in the course of time, relaxation of stress, and development of creep deformation.

Data on the strength specifications of ice, obtained by researchers in many countries as a result of thousands of experiments carried out over the last 100 years, are characterized by a wide range of values; this is because there are numerous factors conditioning the physical properties of ice. For example, the compression strength changes from approximately 4×10^5 to 130×10^5 Pa (or even more), and the bending strength ranges from 3×10^5 to 30×10^5 Pa.

Unfortunately, until recently these results have almost always been presented without sufficient data on the ice structure and the conditions under which various loads were applied: hence it is impossible to use them to determine general rules about the strength properties of ice. The basic result of generalizing these data may be reduced to estimation of the strength of ice as a material if its structure, salinity and temperature are known.

The problem of strength is currently considered from the viewpoint of mechanical and kinetic concepts.

According to the mechanical concept, destruction is a result of loss of stability by a solid body. It is assumed that each material is characterized by a certain threshold stress. When the stress is below that threshold, the body is stable and can retain its integrity under load for an indefinitely long time. This threshold stress is a measure of the strength of the body.

The important thing in the kinetic concept is the <u>development</u> of destruction, which occurs gradually as a consequence of the development and accumulation of submicroscopic cracks. This process develops in a stressed body under the effect of thermal fluctuations. The idea of the lifetime under load, i.e. the time required for the development of the process from the moment of applying a load to a body up to destruction of the latter is significant. It is impossible to give a comprehensive answer to the question of what load a body can endure, i.e. what is its strength, without indication of the time it will be intact. This indicates that the term 'ultimate strength' and 'ultimate rupture stress' are relative. They make little sense in terms of the physical nature of the strength of solid bodies, but are quite convenient in practice.

In considering the mechanical destruction of ice we shall adopt the concept of macroscopic violation of its integrity as a result of various contact effects. Ice destruction is preceded by elastic and plastic deformation. Initial destruction is associated with the development of pores, cracks and other violations of integrity, and complete destruction is characterized by division of the ice body into two or more fragments; brittle failure (without significant plastic deformation), and plastic (or viscous) failure; fatigue; delayed fracture, and other types of mechanical failure.

The fundamentals of the physical theories of fluidity and destruction of solid bodies have developed only in recent years, so it is not always currently possible to explain even quantitatively certain specific details in the destruction of solid bodies in general, and particularly in such a complex material as ice.

The effectiveness (and, consequently, the economic advantageousness) of a particular technique for the destruction of ice is conditioned primarily by major characteristics such as the specific energy of destruction, i.e. the consumption of energy required for the destruction of a unit of volume or mass of ice.

When developing and improving methods for the destruction of ice, it is reasonable from the technical viewpoint to attempt to reduce the specific energy of destruction, thereby causing a decrease in general energy consumption and, as a result, a reduction of the mass and overall dimensions of the equipment required for destruction.

It seems advisable to regard the energy characteristics of ice destruction processes from the point of view of considerations: firstly, the energy approach permits us to analyse methods of mechanical destruction of ice (brittle ice, or ice where progressive creep develops) for the selection of optimal techniques in specific situations. Calculations have indicated that energy consumption for ice destruction by inducing progressive creep is substantially higher (by several orders or magnitude) than energy consumption at brittle failure.

Secondly, considering that it is relatively easy to determine the energy consumption of external background processes (associated for example, with the drift, wind loads or stream), it is possible to estimate their 'projection' on the mechanical behaviour of an ice sheet, i.e. to estimate the energy flux of the effect of ice on a structure and determine its safety factor.

1.1. Perforating and Breaking

Two aspects of this process may be considered, depending on the direction of an impact on one side or the other of an ice sheet: i.e. perforating ice by an impact on the top surface, or breaking an ice sheet by the surfacing of a technical device.

The impact of a solid body on ice and its performation has long been a practical problem. Determination of the laws of interaction of a solid body with an ice sheet on impact is significant in solving problems of dropping a weight, breaking ice by means of an icebreaker, landing aeroplanes, etc. In experiments carried out as far back as 1932 a 57 kg weight was dropped from a height of 2 m and 3 m. Rapidly damping vibration was observed with a period of about 0.1 s, but the results were insufficient to elaborate a method of calculating the ultimate strength of the ice sheet on impact, or perforation. Similar experiments were made in 1967-1969 to investigate the impact of a weight dropped on the ice of Lake Ladoga [86, 162, 163].

Steel cast semispheres with a mass of 300 kg and 156 kg were dropped on the ice of the lake. They carried tensometric overload pickups to record oscillograms of the impact momentum and elastic vibrations in the ice sheet caused by the impact. The contact pressure exceeded 15 MPa, much greater than the ultimate strength of ice on the compression of cubic specimens [86].

The formation of a zone of small-grain ice, separated from the bulk mass by a clearly displayed surface of destruction, and of a zone with substantial shifts in the basic surfaces and irregular displacement at the borders between crystals, are the most essential changes in the ice structure in the contact zone after an impact [163]. The character of these changes in the ice structure in the contact zone is preserved irrespective of the ice temperature regime and initial velocity of the impact within a range of 1-5 m/s.

The specific energy of mechanical disintegration ε_k has been determined, i.e. the energy of destruction referred to a unit of mass of disintegrated substance [162]. The mean value of this parameter, which is comparatively stable for each type of ice, depends on its structure and temperature, changing from 3×10^{-3} J/g for spring ice to 14×10^{-3} J/g for winter ice. The energy of destruction per unit of volume is determined by the formula

$$\varepsilon_v = \sigma_c^2/2E,$$

where σ_c is the respective ice strength at an impact and E is the modulus of normal elasticity.

If we take into account the expression $\varepsilon_v = \varepsilon_k \rho$, for freshwater ice $\varepsilon_v = 3 \times 10^3$ J/m^3.

D.E. Kheisin et al. considered also a mathematical model of forcing a solid body into the ice surface on impact to determine the contact pressure in the zone of impact, which is directly proportional to the specific energy of ice disintegration. N.G. Khrapaty and V.G. Tsuprik carried out experiments to study the performation of ice and appearance of flexural vibrations in the ice sheet on impact [38, 164-169]. Impact loads were created by dropping a 300 kg sphere at a velocity of 1-6 m/s (Figure 1.1).

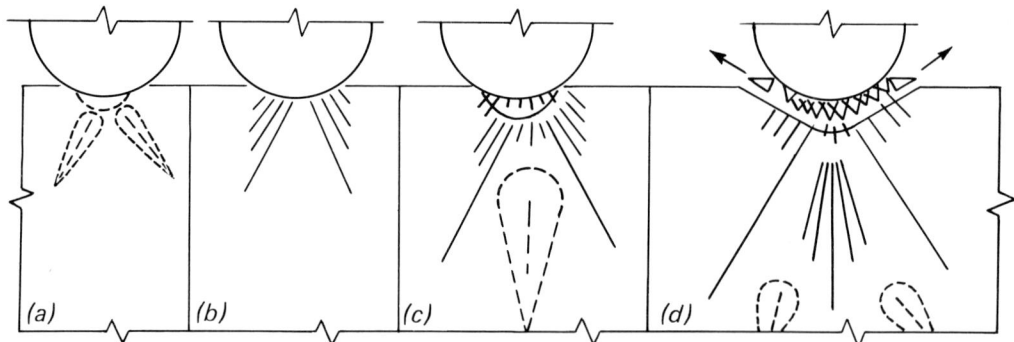

Fig. 1.1. States of ice destruction following the impact of a spherical body on its surface [38].

(a) - elastic sinking; (b),(c) - formation of conical and axial cracks respectively; (d) chipping of lateral cantilevers.

The ice undergoes elastic deformation in the case of a small load, and no visible traces of the impact are observed when the load is removed (Figure 1.1a). An increased load causes the formation and development of concentric cracks that isolate into a cone (Figure 1.1b), which then cracks up in depth.

When the axial cracks reach the lateral surface of the cone, a zone of finely comminuted ice is formed under the contact platform and causes chipping of the lateral cantilevers. A hole forms by the end of the destruction process. The hole is filled with the products of comminution and preserves the remains of the conical and axial cracks (Figure 1.1d). The axial cracks propagate to a greater depth than the zone of fine comminution.

The basic role in this process is played by tensile stress, which causes the formation of conical and radial cracks. These condition the formation of the finely comminuted ice zone. Thus, when a body is forced into ice, compression stress develops under the contact platform in the direction of the load action before the destruction of the ice, and tensile stress develops in directions perpendicular to the action of the force.

The external parts of the cone represent a rigid casing, containing a superstressed internal region in which tensile stress is predominant. The internal superstressed region joins up with the external low-pressure region owing to failure in the casing. As a result, the internal stress is sharply reduced and the liberated component of the potential energy

is converted into the kinetic energy of destruction of the fragments.
   The initiation of cracking at various rates of forcing-in is observed at a constant contact pressure, which is equal to the static ultimate strength of ice specimens at uniaxial compression under experimental conditions.
   Maximum contact stress, which greatly exceeds the static ultimate strength at uniaxial compression, corresponds to the final stage of destruction, i.e. chipping off of the lateral cantilevers.
   Ice may be perforated at the periphery of a load with a high fall velocity and small bearing area, in the absence of snow on the ice surface. The ice strength is limited in this case mainly by its chipping strength. If the ice is not perforated, the load will perform vertical movements after the impact, remaining in contact with the deflecting ice sheet until maximum deflection is reached.
   The results of dynamic and static tests are comparable.
   The dependence of the specific energy of mechanical ice destruction on the temperature and rate of loading and nominal contact stress has been determined. The process of forcing a body into ice has been analysed in terms of the 'force-deformation' coordinates [168]. Investigations have been made of the deflection of an infinite buoyant slab and the time of its occurrence, depending on the force of impact, geometry and elastic characteristics of the slab [169]. A method has been advanced for calculating the depth to which a solid body can be forced into an ice sheet [165]; a criterion has been derived and the specific energy of mechanical ice destruction (SEMID) determined, i.e. the energy which is necessary and sufficient for crushing a unit of ice mass. SEMID is determined quantitatively by the method advanced in [165], as the relation of the irreversibly spent energy of an impact to the ice mass in the volume of the hole formed.
   The specific energy of mechanical ice destruction is a function of its temperature, and is independent of the initial energy of the impact; it therefore has quite a stable value.
   Experiments have been carried out in various parts of the world to investigate the perforation of an ice sheet and vibrations arising therefrom. Spheres of various materials (such as nylon or steel) have been used in the USA [255, 256] to perforate ice at temperatures from 0°C to -20°C. As a result of these experiments data have been obtained on the coefficient of recovery from an impact of a sphere on the surface of a massive ice block. The coefficient of recovery is determined by the relationship of the sphere velocities before and after the collision. The coefficient of recovery decreases with an increase in the impact velocity and an increase of the sphere radius [256]. C.W. Young described an investigation of the parameters affecting the perforating capacity of a radio underwater acoustic buoy [189]. The buoy comprises a striker to perforate the ice, an instruments container and an antenna (Figure 1.2). The striker is designed to perforate ice 3 m thick with a perpendicular impact on the surface at a velocity of 131 m/s. The container penetrates 25 cm into the ice, leaving the antenna on the surface. If the ice is only 30 cm thick, the tip perforates it and goes into the water at a speed of 130 m/s. The container separates from the tip, submerges to a depth of 2 m and surfaces to the ice-water

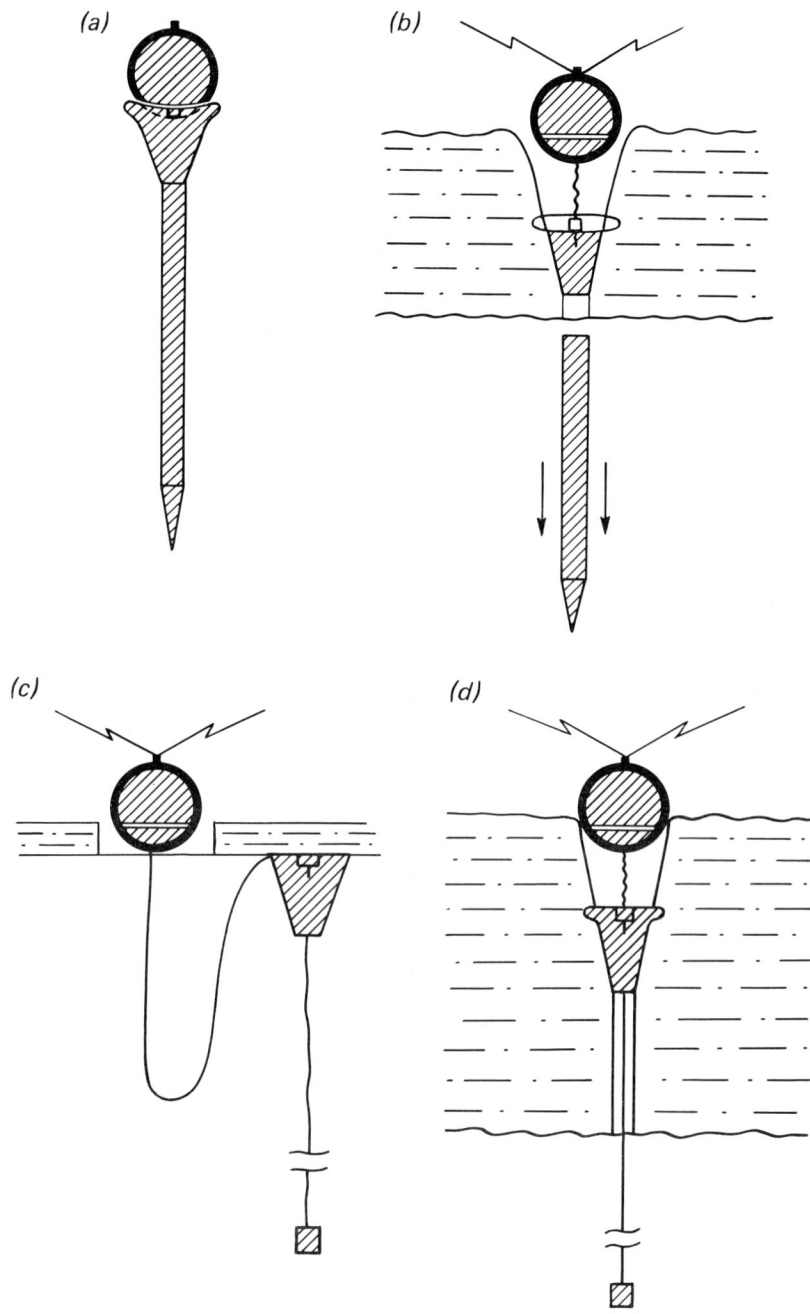

Fig. 1.2. Perforation of ice sheet from beneath by striker of radio underwater acoustic buoy [189].

(a) - configuration in flight; (b) - separation of container and antenna compartment; (c),(d) - perforation of thin and thick ice respectively.

MECHANICAL DESTRUCTION OF ICE

interface. The antenna remains on the water surface in the hole or funnel which forms on perforation of the ice. The tests results demonstrated that the effect of snow cover may be neglected. It has also been determined that the requirements for perforating sea ice and freshwater ice are quite close, although these types of ice differ in their physico-mechanical properties. The optimal shape of the striker is a cone whose length is three times its diameter.

Equations have been derived for calculating the depth of striker perforation into ice, and a coefficient of perforation for the case of ice perforation:

$$D = 3.1 \times 10^{-3} SN(V_i - 100)\sqrt{W/A} \quad \text{for } V_i \geq 200 \text{ ft/s}$$

$$D = 11.7 \times 10^{-3} SN(V_i - 30.5)\sqrt{W/A} \quad \text{for } V_i \geq 61 \text{ ft/s}$$

$$S = \frac{TV_i^2}{(V_i^2 - V_e^2)[3.1 \times 10^{-3} N\sqrt{W/A}(V_i - 100)]} \quad \text{for } V_i \geq 200 \text{ ft/s}$$

$$S = \frac{TV_i^2}{(V_i^2 - V_e^2)[11.7 \times 10^{-3} N\sqrt{W/A}(V_i - 30.5)]} \quad \text{for } V_i \geq 61 \text{ ft/s}$$

where D is the depth of penetration (without perforation), ft (1 ft = 0.305 m), S is the coefficient of perforation (dimensionless), $N = 0.7 \ldots 1.31$ is the coefficient of striker tip serviceability (dimensionless), W is the mass of perforationg device (1b), A is the cross-section area (in^2), V_i is the impact velocity (ft/s), T is the thickness (ft) and V_e is the velocity of the perforating device on egress from ice layer (ft/s).

It is clear from these equations, which are true for $W \geq 22.7$ kg and $W/A \geq 0.35$ kg/cm^2, that the ice thickness and the coefficient of perforation are interrelated values: the maximum ice thickness of 3 m determines the lower limit of the coefficient of perforation, which is equal to 2, while for the softest ice it is 8 (Figure 1.3). The parameters of the impact on ice are considered in detail: impact velocity (91-183 m/s), incline of trajectory, impact angle (between trajectory of device and ice surface), angle of attack. It is shown that the device penetrates into the ice at angles of attack of 40° and rebounds at angles of 36°. A device with a rocket engine, whose cost is naturally higher, has been proposed if it is necessary to develop higher impact velocities than are possible with free-fall acceleration.

The results of three series of experiments on perforation of sea ice by projectiles dropped from an aeroplane, are discussed in a presentation to the International Conference on Engineering Oceanic Environment Research (Ocean-74) [257]. The devices were equipped with

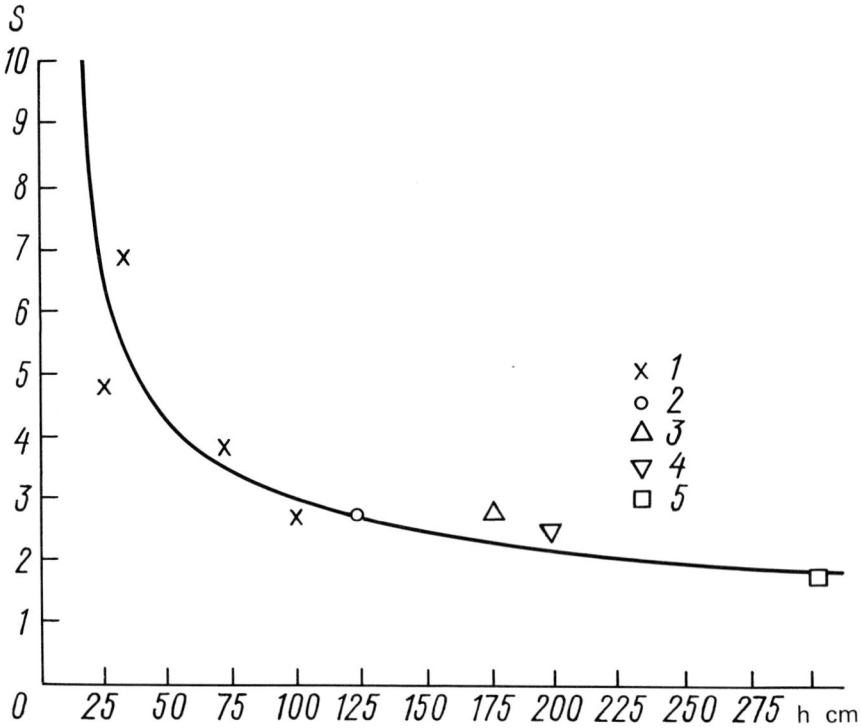

Fig. 1.3. Perforation of ice, depending on its thickness [189].

1 - ice in canals; 2,3 -one-year sea ice (2 - port Clarence, Alaska, 3 - Thule, Greenland); 4 - freshwater ice (Dambell Lake, Canada); 5 - perennial ice.

telemetric means that made it possible to record the acceleration on perforating the ice and information on the parameters of the ocean after perforation. F.I. Legerer, working on mechanics of icebreaking at the Department of Engineering and Applied Science at the University of Saint John's (Newfoundland, Canada), investigated theoretical aspects of ice destruction and produced, in particular, conclusions on the effect of shearing stress under the effect of an impact [227].

A number of technical solutions are realized in devices that destroy ice by numerous impacts on its surface.

Shipbuilders have developed a ship icebreaking system (pat. 3,670,681 USA), comprising two icebreaking members. One of them is arrayed under the ice and is equipped with fork-like teeth which are pressed to the bottom surface of the ice. An adjustable mortising member, breaking the ice, is arranged above the ice surface and lowered vertically downwards (Figure 1.4). Another icebreaking system for vessels patented in the USA (pat. 3,698,340) has a ram, operated by compressed air, arranged on the underwater part of the vessel. The ice

MECHANICAL DESTRUCTION OF ICE

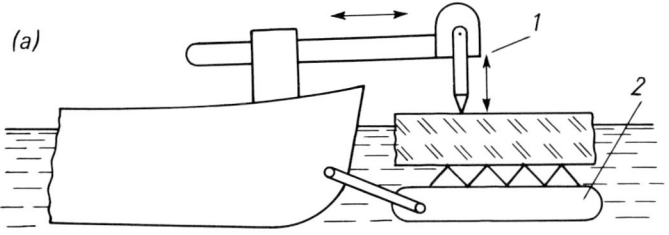

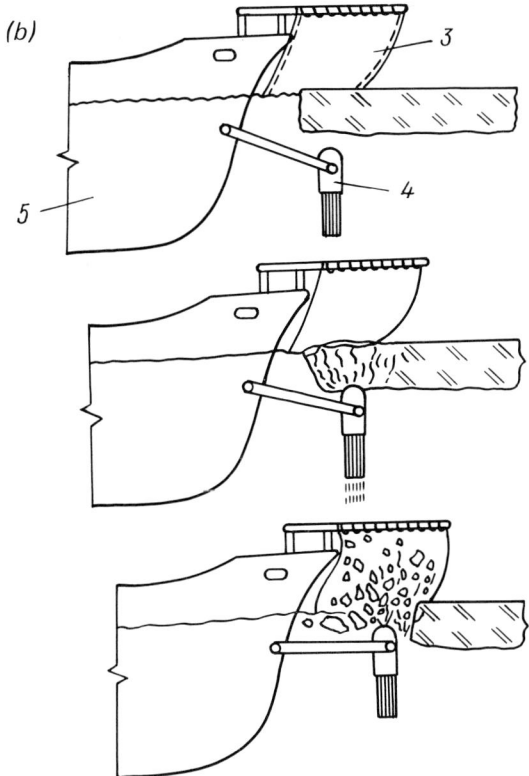

Fig. 1.4. Icebreaker systems for ice destruction by an impact on its top (a) and bottom (b) surfaces [80, 84].

1 - mortising member; 2 - beam with sawlike teeth; 3 - protective screen; 4 - ram; 5 - vessel bow.

is broken by powered upward impacts of the ram with the undersurface of the ice, accompanied by water-hammer at rapid ram oscillation [80, 84].

Analysis of existing icebreaking systems suggests that the

development of the impact method may have many advantages [32].

Penetration of the breaking body into the ice is associated with the creation and development of numerous cracks, depending on the physical properties of the ice and the angle of the body's penetration, its geometrical and technical characteristics. It has been found that there exists a surface of destruction, dependent on the ice properties, wherein the impact development of cracks is ensured by a minimum amount of energy (Figure 1.5). The formation of stress waves of required direction and intensity sets strict requirements for the geometrical and physical parameters of the breaking body, as well as for the wave oscillator, i.e. the vibration or vibro-shock system.

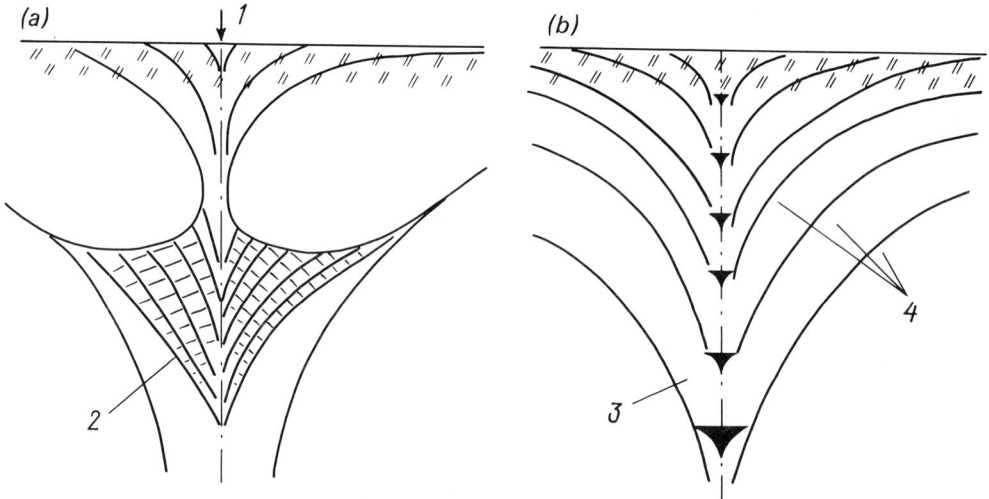

Fig. 1.5. Scheme of (a) ice destruction zones and (b) surfaces (chipped craters), corresponding to minimal energy of destruction [32].

1 - direction of impact; 2 - zone of intensive ice disintegration; 3 - initial angle of chipping; 4 - chipped craters.

A maximum stress gradient at the apex of a crack arises on propagation of an elastic compression wave along only one of its edges. The stress concentration exceeds the ultimate strength of the materials, contributing to rapid propagation of a crack perpendicular to the basic one.

The approach to a crack by a tension pulse at an angle of 90° creates a maximum concentration of tension at its apex. An attack on the crack boundary from a medium with a greater modulus of elasticity causes an increase of stress intensity at its apex. The concentration of stress at an angle of wave incidence is not as great at 180° as it is at 0°.

Elastic stress waves directed across the ice metal interface may cause their mutual lamination, but further disintegration of the ice is

MECHANICAL DESTRUCTION OF ICE 11

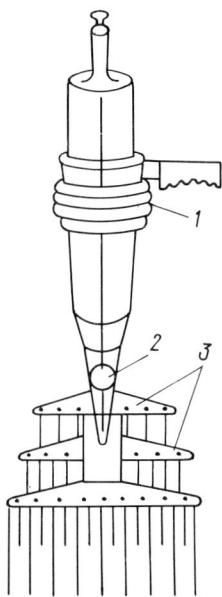

Fig. 1.6. Mechanized hand vibro-shock tool for chipping ice [7].

1- vibro-shock unit; 2 - pneumatic pump; 3 - three-row interferential rake.

bound to be difficult. A three-row interferential rake, enveloped by a pneumatic suction pump and receiving impact energy from the vibro-shock unit, is an ingenious mechanized hand tool (Figure 1.6). The capacity of the tool is 4-6 t/h at a drive power of 0.8 kW.

A large group of inventions deals with vibration devices mounted on vessels. A device in the form of rotating unbalanced masses has been patented in the German Federal Republic (pat. 10,923,336) and in other countries. A vibration unit in the form of a power cylinder with a freely moving mass has been developed in the USSR (Author's Certificate 217,222). These units develop inertial forces, exciting oscillatory motion of the vessel, owing to which the ice sheet is subjected to the intermittent effect of bending moments of rupture. The device is mounted on icebreaker attachments in order to prevent vibration of the vessel proper. An icebreaker attachment with a vibratory system has been developed in the USSR [84] and patented in major Western countries (Author's Certificate 287,532). The attachment is connected to a pushboat by a hinged sliding lock, ensuring linear vertical and angular vertical and horizontal movement of the attachment relative to the pushboat (Figure 1.7). This invention has been utilized in the construction of icebreaker attachments in projects 1713 and 1749 [84]. When testing the attachment on the pushboat RBT-300, ice 45 cm thick was broken at a speed of 0.4 km/h, and, in running tests, ice 70 cm

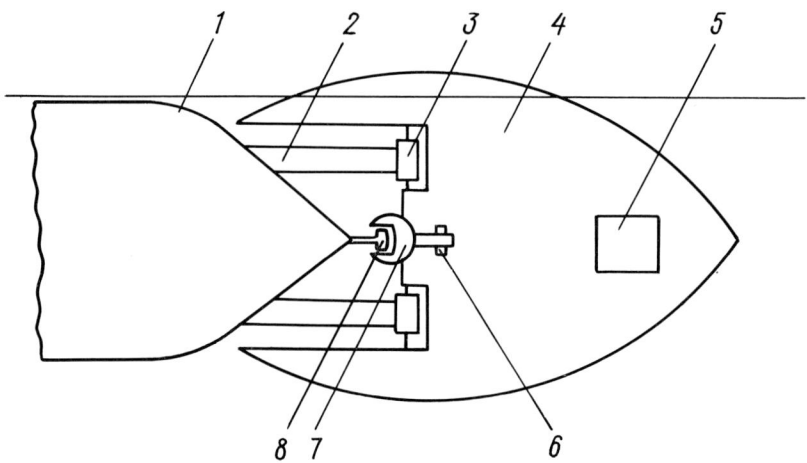

Fig. 1.7. Pushed icebreaker attachment with vibratory system [84].

1 - bow extremity of pushboat; 2 - stops; 3 - freely rotating rollers;
4 - attachment; 5 - vibratory system; 6 - lock coupling axis;
7 - coupling; 8 - rail of hinged sliding lock.

thick was broken at an average speed of 0.1 km/h. The manoeuvring qualities of the convoy tested by turning and chipping ice off vessels, were satisfactory [84].

The experience of utilizing vibration units in river navigation demonstrates that they are an efficient means for increasing the navigational capability of vessels on ice-covered rivers. The maximum thickness of the ice that can be broken is greatly increased if the vibration period of the unit equals the period of natural oscillations of the vessel.

The characteristic feature of this method is disaggregation of ice at low speed. This feature is immaterial in clearing river estuaries of ice, or when removing vessels from a backwater, but it is undesirable in Arctic navigation; vibration units are not therefore found on sea-going icebreakers.

The vibration method of breaking the ice cup on the underwater part of a vessel [31] is based on vibrating the hull by means of a special vibration machine aboard the vessel; for example, a vibratory pile-driver BM-20. The latter method has not been tested sufficiently in practice, and resonance phenomena may induce hazards such as weakening of hull joints and pipe joints, separation of coatings, and possible damage to automatic devices, radio, and other equipment aboard the vessel. Beside that, the parts of the ice cup that have separated from the hull and remain under the vessel's bottom may freeze to it again.

Many patents deal with the method of breaking ice by forces directed upwards under the ice sheet. The broken ice is moved away by the icebreaker, which leaves a clear channel behind. The advantage of this method is the substantial reduction of frictional forces, because the icebreaker stem, breaking the ice from beneath, moves the ice chunks aside so that the port and starboard sides of the vessel are not in contact with the ice.

This method has been developed in Canada and the German Federal Republic in several variants. The Canadian company Alexbow filed three patents for different designs of an icebreaker stem designed to break ice from beneath upwards. One of these patents (pat. 1,267,079 Canada) envisages the use of a bulb-shaped bow extremity. The 'Canadian bulb' is characterized by considerable width, and the height of the bow's 'cutter' is 3/4 that of the ship's draught. Tests have demonstrated that the bulb ensures a gain in speed of 0.4-0.5 knots, and the friction coefficient of the ice against the steel in the underwater position is reduced to 0.01. Also, adhesion of the ice to the hull is eliminated.

Similar designs have been patented in England (pat. 1,215,529 and 1,215,530), the USA (pat. 3,521,590), France (pat. 1,577,665), Japan (pat. 33901/71), and the USSR (pat. 315,341).

The task of improving the navigation of a vessel in ice, for instance by reducing the forces required for breaking the ice and ensuring an ice-clear channel, may also be solved by modifying the traditional geometry of the stem and the shape of the hull. It is assumed that the ice thickness will be greater and its fragments will not go under the hull, thereby improving the navigational capability of a vessel in ice [80].

The West German variant of an icebreaker vessel's hull (pat. 2,206,472 FRG) is distinguished by the special lines of the hull (Figure 1.8). The underwater bulb, inclined downwards, has a rounded

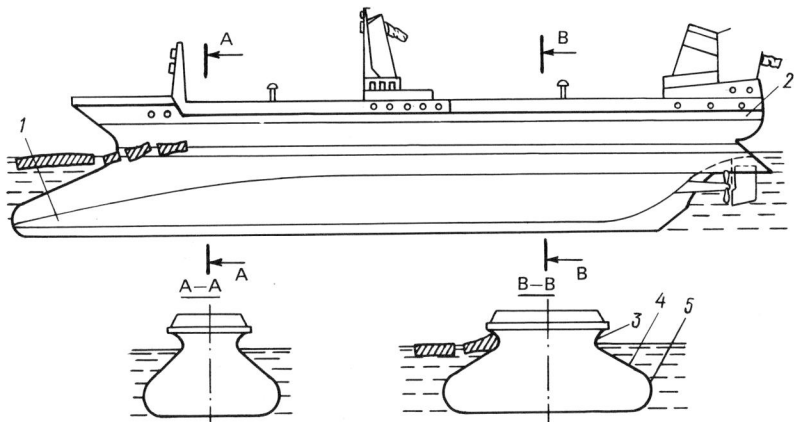

Fig. 1.8. Breaking ice from beneath by underwater bulb of icebreaker with a variable dip angle of sides [80].

1 - underwater bulb; 2 - guard of screw and rudder; 3 - ice strake; 4 - inclined side chamfers; 5 - convex underwater part of vessel.

central part. The operative waterline is on the inclined side chamfers. The ice is broken from beneath by the pressure of the bulb upper surface and is moved aside by the chamfers, which end in the stern part of the hull with a guard to protect the screw and rudder. The ice strake of the outside plating is concave, turning gradually at the bottom into the chamfers. The strake prevents the advance of broken ice onto the upper deck. The convex part of the hull reduces the areas of the operative waterline, thereby reducing ice resistance to navigation of the vessel and improving its manoeuverability [80, 85].

An icebreaking attachment has been developed to break river ice (pat. 2,229,621 FRG). The attachment (Figure 1.9) is a pontoon braced to the pushboat. The pontoon is trihedral in plan, and its bow extremity is a pear-shaped bulb. When the vessel moves, the bulb develops a hydrostatic wave which breaks the ice sheet from beneath, and the horizontal forces required for breaking the ice are reduced.

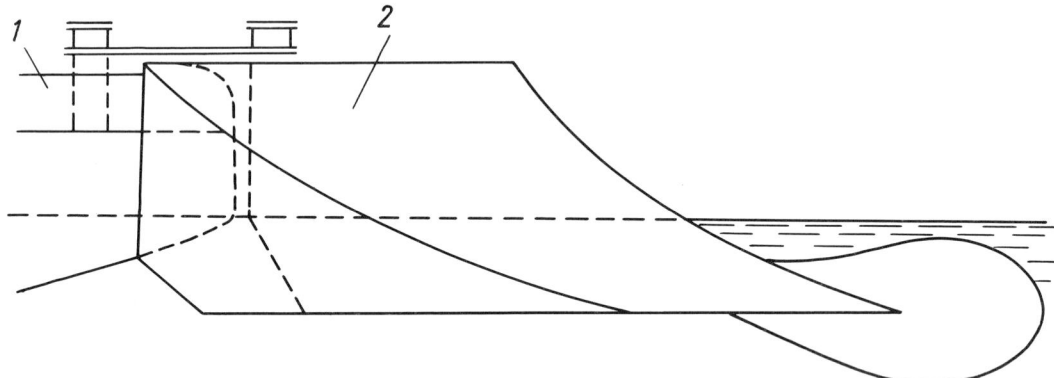

Fig. 1.9. Icebreaking attachment for breaking ice from beneath [84].

1 - pushboat; 2 - pontoon.

One of the mechanical methods of ice destruction is based on the principle of eliminating the elastic hydraulic base, i.e. water from beneath the ice (pat. 3,808,997 USA). An icebreaking arrangement supplies air under sufficient pressure to lower the surface of the water to a level below the bottom of the ice. The unsupported weight of the ice induces mechanical failure and it breaks off. According to another method (pat. 3,841,252 USA), the hydrostatic upthrust is eliminated by injecting gas beneath the ice sheet and breaking it by applying force. A largely new device (pat. 3,572,273 USA) for breaking ice has been developed at the Southwest Research Institute in San Antonio, Texas [141]. The bow of a self-propelled barge is fitted with a rectangular metal pontoon, having in its front part a large combustion cylinder of an 'internal combustion engine'. The ice floe is the 'piston' of this cylinder. The barge approaches the ice sheet and positions the pontoon beneath it. The arrangement operates as follows (Figure 1.10). An air-propane mixture (ratio 30:1) under a pressure of 0.4 MPa is

MECHANICAL DESTRUCTION OF ICE

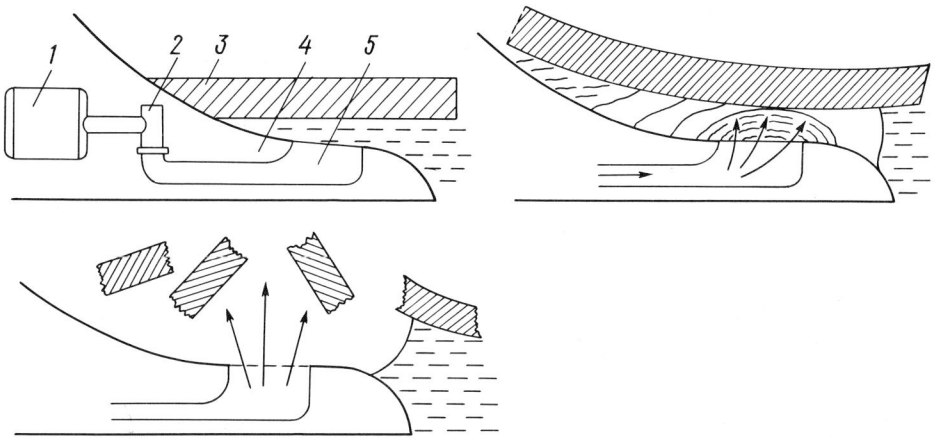

Fig. 1.10. Ice destruction by air bubble of underwater gas exhaust [245].

1 - combustion chamber; 2 - quick-acting exhaust valve; 3 - ice sheet;
4 - inclined platform passing beneath ice sheet; 5 - gas exhaust opening.

injected into the combusion chamber and ignited by a glow plug. The exhaust valve opens when the pressure in the chamber increases by a factor of 6-8. Compressed gas is discharged with an explosion via an exhaust pipe, creating an air cushion which lifts and breaks the ice. The successive stages of icebreaking are shown in Figure 1.10. A similar arrangement, producing repetitive combustive explosions every 10 s, was designed for the protection of buoyant structures and offshore drilling derricks against ice [245].

A mounted arrangement may be used on stationary hydraulic structures. The basic element of this arrangement, which produces an underwater controlled explosion of a gaseous mixture, is suspended on a bracket by means of a hinge to prevent the transmission of force from the explosion to the suspended equipment (Figure 1.11).

A system using a new icebreaking principle has been advanced in the USA for an icebreaker attachment which looks like an 'ice plough' (Figure 1.12), and is not connected rigidly to the hull of the vessel. The attachment comprises a combustion chamber, operating on hydrocarbon fuel. High-pressure gas is discharged intermittently into the water under the ice. The system has been tested under natural conditions. The pressure in each of the three 0.14 m^3 combustion chambers was 2.8-4.2 MPa. Ice 30 cm thick was broken across an area of 9.3 m^2 at each discharge of gas. The US Coast Guard, who financed the development of this system, planned to design one that would clear a channel 12.2 m wide in ice 61 cm thick at a speed of 9 km/h. Its operation would require an additional 300-400 kW [249].

Efficient methods of increasing the navigational capacity of vessels in ice are based on reducing the frictional force between the

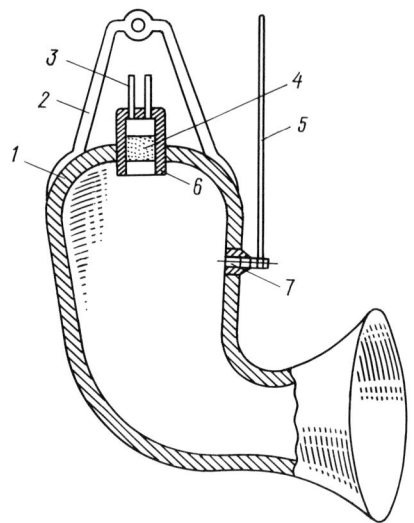 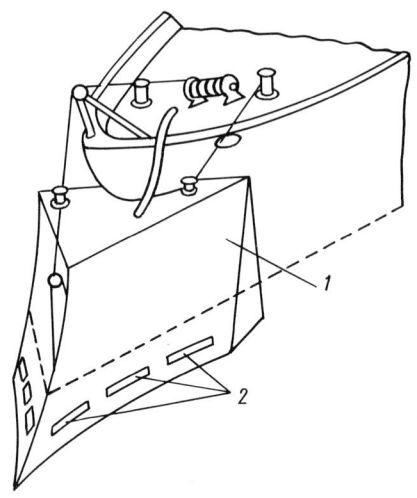

Fig. 1.11. Arrangement for breaking ice by a standing wave from an underwater gas exhaust [141].

1 - body; 2 - suspension;
3 - pipes supplying combustible mixture; 4 - antiknock agent;
5 - power supply; 6 - outlet;
7 - glow plug.

Fig. 1.12. Icebreaker system for breaking ice by repetitive underwater gas exhausts [249].

1 - plough-shaped body of system;
2 - exhaust openings.

hull and the ice. An arrangement designed to reduce the friction coefficient between the ice and the hull of a vessel by means of air bubbles discharged along the hull and creating an effective air lubricant, has been developed by the Finnish Vyartsila shipbuilding company [61, 221].

Canadian shipbuilders have designed an icebreaker bulker with a 28000 t deadweight and a speed of 15.5 knots. The air-bubble chamber reduces friction of the hull against the ice, while variable-pitch screws increase the thrust at low speeds.

The development of oil and gas deposits on the bottom of ice-covered Arctic seas has made it necessary to study the deflection and destruction of ice sheets caused by a mass of gas that forms as a result of an oil or gas blowout.

D.R. Tophman, of the sea ice research group of the Oceanic Environment Control Department (Victoria, Canada), advanced an analytical solution for deflecting an infinite ice sheet in the centre and at the edges of an underwater bubble of gas, and tested it on a model [250].

Simulation results have demonstrated that the destruction of an ice sheet may occur at the centre of the gas bubble or above its edge,

MECHANICAL DESTRUCTION OF ICE

depending on the thickness of the ice and the amount of gas as well as on the ice properties. Breaking of an ice sheet 1 m thick by a gas bubble with a thickness more than 100 mm will probably occur at the periphery of the bubble. The critical radius of the bubble at which ice destruction occurs depends on its thickness and on the physico-mechanical properties of the ice.

Water incompressibility is the basic feature in breaking ice by a gas bubble, because it ensures that pressure developing at the water-ice interface is transmitted to the ice sheet. Ice is characterized by comparatively low resistance to bending stress, so the application of such methods for the destruction of ice seems quite logical.

An icebreaker has been patented in Denmark (pat. 95,983), whose principle of operation is as follows. A body of variable buoyancy is submerged under the ice; it has a maximum upthrust force sufficient to lift and break the ice. The icebreaker (Figure 1.13) comprises a top

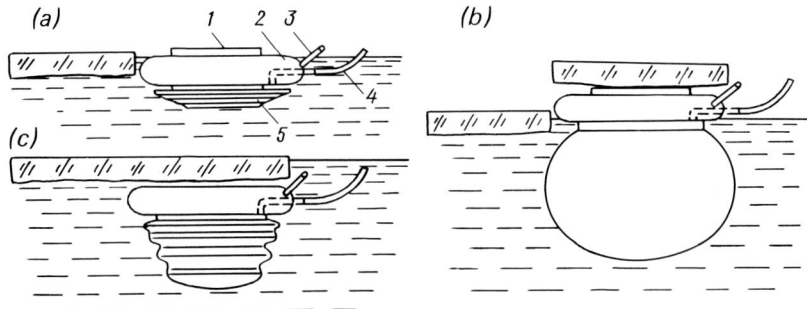

Fig. 1.13. Mobile icebreaker [141].

a,b,c - icebreaking stages; 1 - stiffening rib; 2 - top pontoon; 3,4 - outlet and inlet pipes; 5 - bottom pontoon.

pontoon of constant volume, having an elongated shape with a slightly lowered bow and an inflatable lower member, which is fixed under the top pontoon and is connected to a compressor plant by a hose. The upper and lower elements are connected to a leading vessel which has a compressor or a pump to remove the ballast.

Before starting operation, the system has sufficient buoyancy to stay on the surface. When it is sunk under the ice, compressed air is supplied into the bottom member. This increases the buoyancy of the icebreaker to the maximum, and the ice breaks above it. The gas is then discharged, and the operation is repeated. The exhausted gas may be used to propel a jet, and the top part of the icebreaker may be made rigid (steel or reinforced rubber) and equipped with rakes or knives to facilitate the icebreaking operation. The icebreaker may operate automatically if an automatic charge valve and a device riggered by a certain contact force of the apparatus with the ice above it are built in. The icebreaker may be constructed on the principle of variable

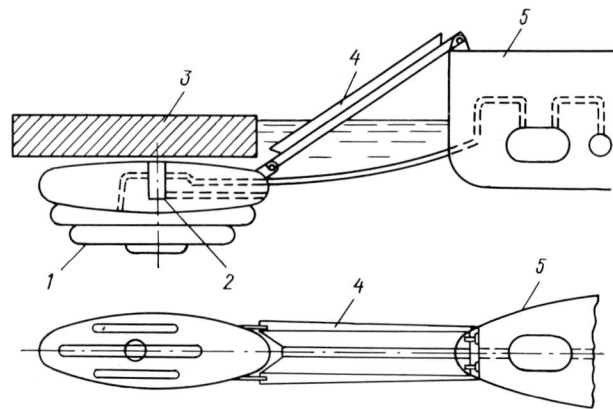

Fig. 1.14. Submersible icebreaking apparatus [80].

1 - submersible icebreaker; 2 - buoyancy system; 3 - ice sheet; 4 - cantilever; 5 - prow extremity of vessel.

buoyancy, by supplying and pumping out water ballast.

A similar submersible device fitted in the prow of a vessel (Figure 1.13) has been patented in the USA (pat. 3,130,701).

The method of breaking ice by air bubbles contained in submersible flexible casings is quite effective. When compressed air is supplied, the casings surface and act on the ice with a force equal to the weight of displaced water. A 50 m^3 casing develops an upthrust force of 500 kN. The radius of the broken lane greatly exceeds the dimensions of the casing [14].

The system consists of flexible casings with necks and flexible hoses (Figure 1.15). The casings are connected to a compressed air source and are interconnected by a flexible pipeline. The casings may be connected to a displacing mechanism; for example, by a rope to a winch. The air is discharged from the casings when the ice is broken, and the casings submerge into the water body by gravity. The process can be accelerated by a vacuum pump. The flexible casings are moved to a new position by means of a rope, and the cycle is repeated.

The latter system is more preferable in some cases to the systems described above (for example, to increase the effectiveness of dredges under winter conditions), owing to its simplicity and relatively low cost.

Hoverships have been used as a means for ice destruction only since 1972, when, on towing an air-cushioned icebreaking attachment ACT-100 at a speed of 6.4 km/h across the Great Slave Lake at the mouth of the Yellowknife River in Canada, it was found that a channel remained in ice 68 cm thick after passage of the attachment [252].

A light icebreaker, the Alexander Henry, with the same ACT-100 attachment on an air cushion, was tested in 1972 in Thunder Bay, Ontario,

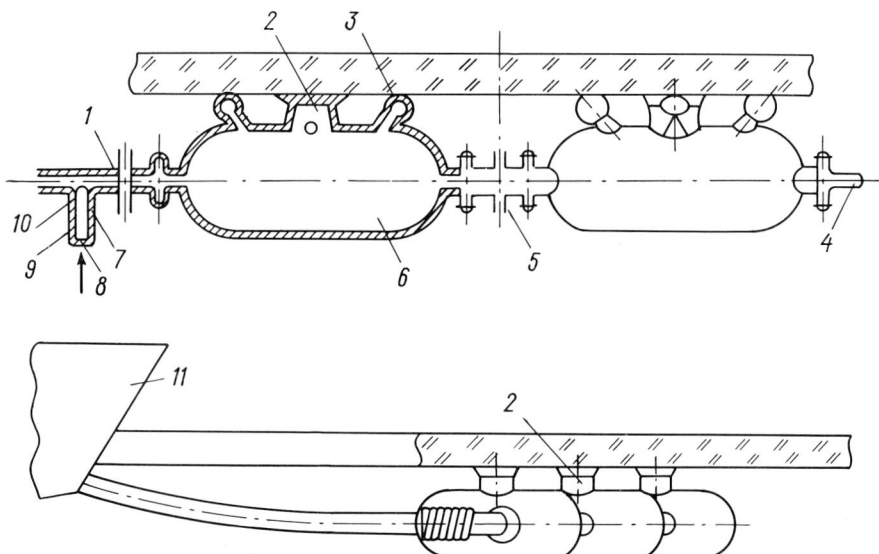

Fig. 1.15. Design and connection of flexible casings to break ice by compressed air [14].

1, 5 - flexible pipelines; 2 - necks; 3 - flexible hoses; 4 - rope to winch; 6 - flexible casings; 7, 8 - conduits; 9 - piston; 10 - control check valve; 11 - vessel's hull.

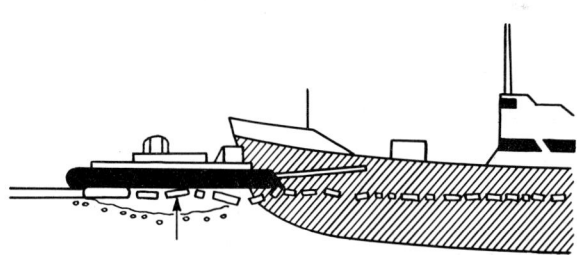

Fig. 1.16. Air-cushioned attachment.

Canada (Figure 1.16). The speed of the vessel without the attachment was 3.7 km/h in the case of 36.5 cm ice, and 16.7, 9.2 and 3.7 km/h with the attachment fitted at ice thicknesses of 43, 60 and 78 cm respectively [45, 138, 197]. The platform broke up ice 40 cm thick when moving ahead of the icebreaker at a speed of 12 km/h during tests in Thunder Bay in 1975 [212]. An air-cushioned means of transport has also been tested on the ice of Antarctica [211].

Modernization of the ACT-100 platform has made it possible to

design the so-called 'ice-eater' [179, 220]. A V-shaped recess is cut
in the stern of the platform for the vessel's bow. According to the
opinion of some specialists, the unsupported ice sheet is broken under
the platform when the air pressure moves the water away from beneath
it [196, 220, 222, 239].

The hovercraft Voyager was tested in the spring and summer of 1974
in the Arctic, and cruised 1600 miles in the rivers of Alaska and offshore regions of the Arctic Ocean [192, 233, 253]. The ship was equipped
with two 250 kW gas turbines, the air cushion pressure reaching 24 kPa.
The cost of this hovercraft was $1.25 million. The Voyager broke ice
blocking the Rivière des Prairies (north of Montreal, Canada) at a
speed of 1 mile/12 h. The ice thickness was 45 cm. The icebreaking
regime was as follows. First, the hovercraft developed a series of waves
in shallow water, cruising at a speed of 10 knots. The speed was then
reduced so that one of the waves would precede the ship in meeting pack
ice. Repeating this manoeuvre many times, the hovercraft broke up the
ice field into chunks of ice (Figure 1.17). The Voyager broke thinner
ice at a speed of about 15 knots, moving in a zigzag across the cracks
on its surface [192, 233, 239, 254].

Additional tests, carried out in the same month in Toronto, Canada,
demonstrated that the amplitude of the standing wave at a maximum speed
of 15 knots was 1.6 m and the ice broke continuously, though its thickness was 38 cm. The width of the channel astern was greater than at low

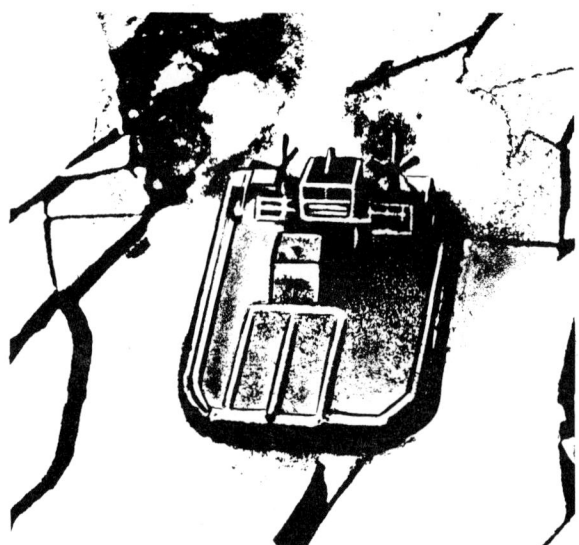

Fig. 1.17. Destruction of ice
field by the hovercraft
Voyager.

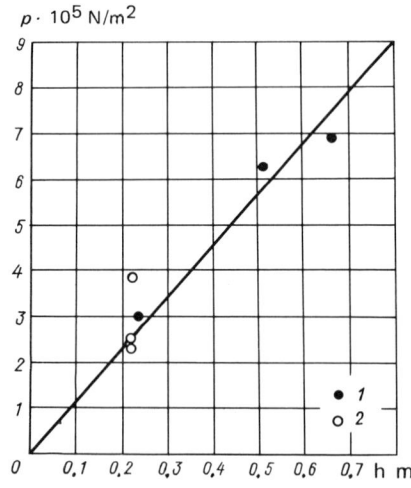

Fig. 1.18. Dependence of ice thickness on pressure of ice cushion
beneath air-cushioned platform and
hovercraft [62].

1 - air-cushioned platform ACT-100;
2 - hovercraft Voyager.

speeds, but the broken floes were big. Modern technology makes it possible to use a cushion pressure equivalent to a water column pressure of 135 kPa [252]. Two methods of icebreaking have been tried out. In the first method, a hovercraft cruising at a low speed supplies compressed air under the ice, which then breaks because it is deprived of an elastic hydraulic base. The thickness of ice, broken at a speed up to 6.5 knots, was 40 cm [254]. According to other data, the hovercraft broke ice 80 cm thick at a speed of 4 knots. In the second method, ice was broken by a wave system astern the hovercraft, cruising at a speed of 13.5 knots [204].

Summary data on the results of breaking an ice sheet by a hovercraft are given in Figure 1.18 and Table 1.1.

Table 1.1.
Results of breaking an ice sheet by a hovercraft

Tactical characteristics				Parameters of ice sheet			
Type or name of air-cushioned platform (hovercraft)	Mass of air-cushioned platform (hovercraft) (10^3 kg)	Propulsion power (kW)	Pressure in air cusion (10^5 N/m^2)	Thickness (m)	Strength (10^5 N/m^2)	Load-carrying capacity (10^4 N)	Modulus of elasticity (10^5 N/m^2)
ACT-100	263	941	0.07	0.68	8.79	193.4	35100
H-119	13.4	144	0.027	0.23	7.03	24.2	24600
HJ-15	16.5	262	0.030	0.25	-	-	-
Voyager	40.8	1912	0.025	0.23	7.03	41.3	24600

According to R.G. Wade, Secretary of the Canadian Special Interdepartmental Commission for studying the potentiality of breaking ice by hovercraft, the maximum thickness of ice that can be broken by this vessel at a speed of 20 knots is 1 m. A 15 km^2 ice sheet can be broken in 1 h [137, 252].

The theme of icebreaking by means of hovercraft occupies a prominent place in the long-term plan of scientific research on ice engineering in Canada (as a consequence of results of numerous natural field tests, simulation and theoretical investigation of icebreaking). The expected result is an estimate of the economic and technological potentialities of utilizing hovercraft for icebreaking [253], as well as recommendations for optimal manoeuvring of a vessel to attain maximum icebreaking effect [89, 252].

The department of navigation safety of the Canadian Ministry of Transport carried out experiments on the potentiality of using icebreaking air-cushioned devices and hovercraft at a wide range of cruising speeds, air cushion pressures and ice thicknesses. As stated above, the tests demonstrated two trends in the potential use of hovercraft as icebreakers: the use of low or high speeds.

A non-propelled air-cushioned platform was used as a low-speed icebreaking means, and the hovercraft Voyager as a high-speed means.

Icebreaking occurred at low cruising speeds (up to 4-5 knots) owing to static bending of the ice sheet by the pressure in the air cushion at the edge of the ice sheet.

The physics of ice sheet breaking at low speeds can be described as follows. During motion, the pressure in the cushion lowers the water level beneath the hovercraft under the bottom surface of the ice sheet. An air cavity, whose dimensions exceed those of the hovercraft, is formed under the ice. The ice sheet overhangs the water in cantilever and breaks by gravity when the length of the ice cantilever reaches the critical value (Figure 1.19).

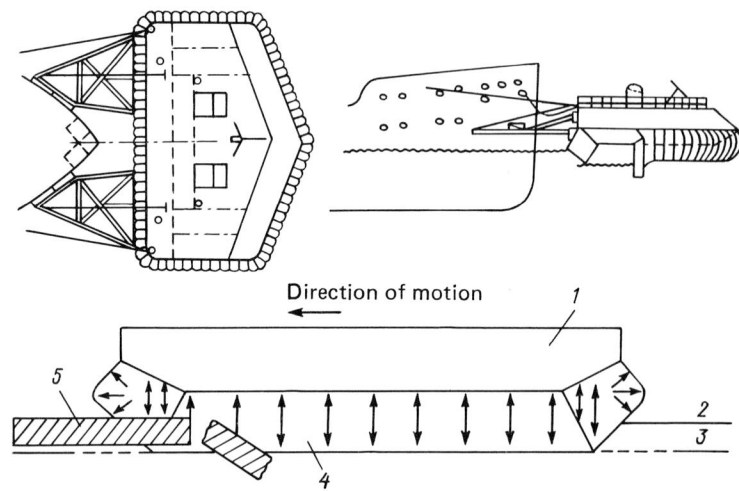

Fig. 1.19. Interaction of air-cushioned arrangement with ice sheet [179].

1 - air-cushioned platform; 2 - water level; 3 - depression in water; 4 - air cushion; 5 - ice. The direction of motion is indicated.

The ice breaks at high hovercraft speeds (up to 10-15 knots) as a result of flexural vibrations under the effect of the moving pressure source, causing wave strain of the ice and water surface.

The maximum ice thickness that may be broken by a hovercraft under a static regime is proportional to the depression depth under the vessel in pure water and inversely proportional to the ice flexural strength. The maximal ice thickness broken by a hovercraft is not its only characteristic: the criterion for estimating hovercraft effectiveness is expressed by the formula

$$k = N/(h^2 V b_c C_c)$$

where k is the criterion for estimating effectiveness, N is the vessel's power, h is the ice thickness, V is the vessel speed, b_c is the channel width and C_c is the coefficient of channel purity (which varies from 1 to 0).

When the hovercraft is at its critical speed, the width of the channel is the width of the zone in which the ice is broken by the air cushion [62].

Ice which freezes up on vessels and hydraulic structures is broken off by means of anti-icing systems.

Pneumatic anti-icing systems are protectors (usually of rubber) that adhere tightly to the surface of the object to be protected. When the anti-icer is engaged, the protector is inflated by compressed air and the ice formed is broken off. Pneumatic anti-icing systems are generally used to protect the wings and stabilizer of propeller-driven aeroplanes. The anti-icer produced by the Palmer company, designed to protect the front edge of a wing, weighs 30-35 kg, and its consumption of compressed air is 0.4 kg/min.

Pneumatic anti-icing systems comprise mobile casings made of polyethylene, polyurethane, etc. (pat. 1,354,875 GB), frost-resistant elastic rubber (pat. 3,744,690 USA, pat. 2,091,955 France, pat. 1,331,698 GB).

The use of silicoorganic, fluorine and reinforced frost-resistant rubber makes it possible to reduce greatly the mass and raise the effectiveness and reliability of pneumatic anti-icing systems [134].

Certain pneumatic anti-icing devices have been tested under sea conditions. For example, a pneumatic anti-icing device, actuated by compressed air, was tested on a British trawler Boston Phantom near Iceland in 1969. This was a neoprene cover fitted on the masts, shrouds, stays, and face walls of deckhouses and superstructures. The cover contained small-diameter rubber tubes that were connected to a central air source supplying air at 0.105 MPa. When the air entered the tubes the cover was inflated, and the ice frozen on its surface dropped off. The system was remotely controlled from the bridge. It is impossible to protect all the surfaces on a vessel by this means; for example, it is difficult to use it in order to protect the trawl winch. Laboratory tests have demonstrated that a layer of ice 23 mm thick formed by atmospheric icing is broken and removed from the model of a mast and the face wall of a deckhouse at -15°C.

The pneumatic method is best suited to removing ice of atmospheric origin. According to V.V. Panov [120], it is more difficult to remove sea splash icing from a surface by means of a pneumatic anti-icing system because of the viscosity and porosity of the ice. Hence, although they are widely advertised, mobile casings are not widespread on seacraft.

1.2. Cutting, Milling and Chipping

Ice-cutting machines of different types are used for breaking ice in areas where icebreakers cannot operate in shallow waters, or in combination with other methods. Ice-cutting machines are divided into three types: ice saws (with chains or jibs), ice cutting machines, and ice ploughs.

It has been established that the destruction of ice is of a fragile character when ice-cutting machines operate with end-milling cutters [97]. The reasons for the occurrence and development of fragile destruction are the accumulation of dislocations preferentially at the boundaries of the grain, conditioning stress concentration at these spots and causing the occurrence of microshears in the crystal, which, in their turn, cause microcracks. The cracks do not propagate linearly because they reflect from obstacles in the form of anomalies of wave resistance of the envirobment.

The real potentiality of crack formation is conditioned by the level of local overstress and the number of simultaneously occurring shears. Therefore, plastic deformation in ice at maximal stress causes the formation of cracks whose sizes increase rapidly to the critical level. If the sliding planes are perpendicular to the direction of compression, shear and crack formation are impeded. According to Payton [241], when the direction of sea ice compression is changed the compression strength increases four times (from 2 to 8 MPa). Lavrov [78] came to the same conclusions (Table 1.2.).

Table 1.2.

Shear strength of polycrystalline freshwater ice at -3°C, depending on direction of force [78].

Sample dimensions			Speed of deformation (cm/s)	Loading time (s)	Shear strength (MPa)	Loading conditions
Width (cm)	Thickness (cm)	Cross-section area (cm^2)				
10.6	10.7	113	0.06	3.5	0.34	
10.9	10.0	110	0.06	2.0	0.72	
10.5	9.0	95	0.06	2.0	1.22	

The dark strip in the figures in the table designates conditionally the metal strap used to supply pressure to the ice. It is evident that the shear strength of the ice along the optical axes, shown by thin lines, is much less than at shear tests across the axes.

It is interesting to compare the operation of ice-cutting machines with end- and side-milling cutters [185]. At the initial moment, when the chip thickness is small, a side-milling cutter directs the cutting force across the crystals (Figure 1.20). When the chip thickness reaches

MECHANICAL DESTRUCTION OF ICE 25

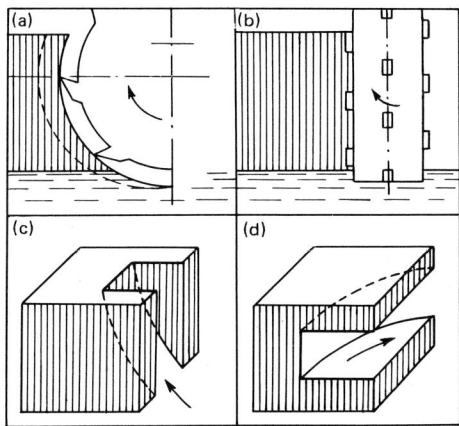

Fig. 1.20. Cutting direction in relation to orientation of ice crystals [185].

(a),(c) - for side-milling cutter; (b),(d) - for end-milling cutter.

maximal values, cutting proceeds in the direction which is most favourable from the viewpoint of ice resistance (σ_{shear} = 0.34 MPa). Cuts of rational sizes and periodicity reduce the cutting force and facilitate advance of the cutter to the free ice surface, thereby reducing load pulsation on the cutter and in the parts of the ice-cutting machine's transmission (Figure 1.21). An end-milling cutter is used to cut in transverse planes; in this case the cutting force is directed during the entire cutting cycle in accordance with the second

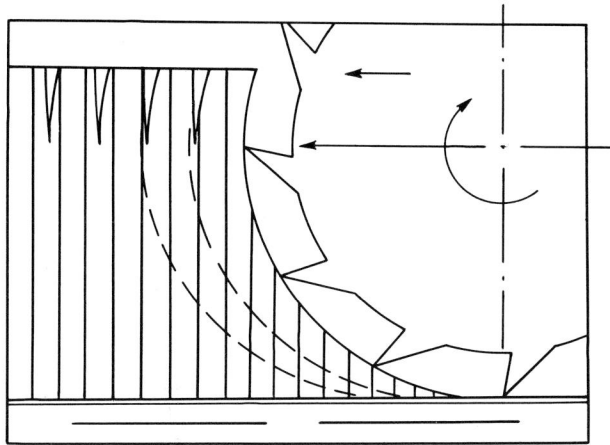

Fig. 1.21. Cutting ice with artificial transverse cuts made by side-milling cutter [185].

case of loading as given in Table 1.2 (σ_{shear} = 0.72 MPa). Therefore, when an end milling cutter is used, the ice-cutting resistance is much greater than in the case of a side-milling cutter. It is also necessary to consider the manifestation of anisotropy in the mechanical properties of ice when designing and operating ice-cutting machines.

1.2.1. Cutting Means

Ice saws with chains are simple in design and have high capacity. Several such machines have been designed on the basis of the petrol saw Druzhba (Friendship), e.g. models DLN-1 and LM-3. However, practical experience has demonstrated that the narrow cuts sawn by these machines in ice close up again rapidly, and the frozen ice recovers its initial resistance to an applied load at very low subzero air temperatures.

The ice-cutting machine designed by N.F. Kharlamov is a mechanism of this type 147 . It is a mobile non-propelled unit fitted on skis and comprising a petrol engine of the Druzhba type, a transverse brace, and a sawing unit with a chain tensioned on sprockets. The mass of the machine is 30 kg. The sawing unit may incline around the horizontal axis and cut ice without preliminary preparation. The testing of this machine was reassuring. It made a rectangular ice-hole instead of a round one, and the ice prism, whose height was equal to the ice thickness, was extracted to the surface. The major advantage of the machine was its mobility.

V.F. Ovchinnikov [148] carried out a wide range of tests to study the process of cutting ice with chain saws in order to investigate the physical essence of ice cutting, substantiate the cutter geometry, and determine optimal cutting regimes. The power used when sawing ice with a chain saw unit was calculated by the following formula.

$$N = k_1 P V_c / 102 \eta$$

where N is the power (kW), k_1 = 1.5...2.0 is a coefficient taking into account brief overloads that occur when the ice chips enter the space between the chain and the teeth of the sprockets, ice chip grain enter the space between the lateral chain links and the saw cut walls, and other factors, $\underline{P}$ is the circumferential force on the saw chain drive sprocket, depending on the technological parameters and physicochemical properties of the ice (kg), $\underline{V_c}$ is the cutting speed (m/s) and η is the efficiency of transmission from the engine to the saw chain drive sprockets taking into consideration losses occurring when the chain bends around the sprocket.

Investigations of the speed of crack propagation ahead of the tooth cutting edge are of interest from the viewpoint of the physical process of cutting. It may be assumed that, if the cutting speed exceeds the speed of crack propagation, the expenditure of power in the cutting process exceeds the optimal value. But if the cutting speed is below the speed of crack propagation, the cutting speed and the machine capacity respectively are below the optimal values although the

expenditure of power for cutting is low. It is necessary to determine
the optimal cutting speed at which the expenditure of power is minimal
at an optimal value of cutting speed and machine capacity. Original
methods and high-speed filming have been used to determine the speed
of crack propagation in ice, depending on various loading conditions:
from several metres per second to speeds of shear waves in ice [16, 147].

The ice-cutting machine developed on the basis of investigations
by V.F. Ovchinnikov (Figure 1.22) can make square ice holes in an ice

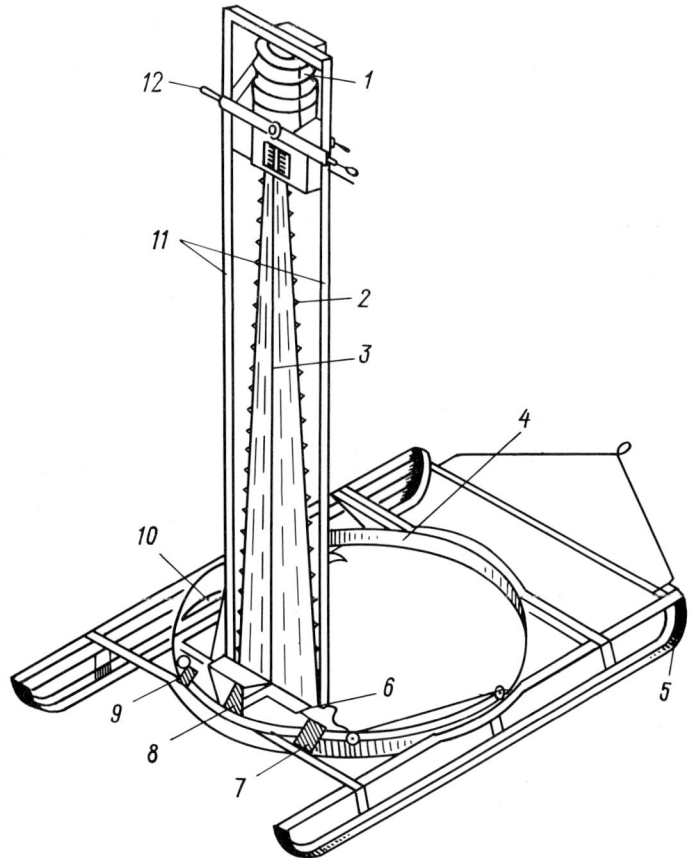

Fig. 1.22. Ice-cutting machine for square lanes [148].

1 - petrol engine; 2 - saw chain; 3 - saw blade; 4 - guide rim;
5 - sledge; 6 - supporting frame; 7, 8, 9 - cutout pedals of posts,
guide fork, rotary sector respectively; 10 - rotary sector; 11 - guide
posts; 12 - feed handle.

sheet with a thickness up to 120 cm and subsequently remove the ice
prisms from the holes. The power of the petrol engine at nominal
rotational speed is 3.2 kW, and its service life is 500 h. The cutting
speed is 4.5-5.2 m/s. The machine is brought to the site of operation,
the engine is engaged, and the saw blade is directed downward. The saw

chain begins to cut the ice, penetrating gradually into the cut. When
the bottom edge of the saw blade reaches the water, it is lifted along
guides to the initial uppermost position. The rotary sector is then
turned through 90° and the next cut is made. Two more cuts are made
in a similar manner. The machine is then moved aside and the ice prism
is extracted with the aid of hooks. The machine saws a 35 × 35 cm^2 ice-
hole at an ice thickness of 0.6-1.0 m in 1-1.5 min with a petrol
consumption of 40-50 g. One tank of fuel is enough to make 30 ice holes
if the engine's fuel tank capacity is 1.5. The length of the machine
(sledge length) is 130 cm, its width 80 cm, its height 168 cm and its
mass is about 40 kg. This machine has been put into regular production.

According to data in [14], ice cutting by means of a jig actuator
is characterized by relatively low power consumption and traction force,
making it possible to design small ice-cutting systems with unsinkable
sledges. Indeed, according to operating data, the minimal value of power
consumed is 12.9 kW at a chain speed of 1.3-1.5 m/s when cutting ice
0.95-1.0 m thick at an air temperature from -12° to -24°C with a jib
actuator. The required traction force does not exceed 22 N at a feed
rate of 120 m/s. When designing chain ice-cutting machines, it is
advisable to incorporate anticutting from the top downwards to remove
the chippings to the water. The potential rapid repeated freezing of
the cut filled with surfacing chippings is eliminated by fitting augers
on the axle of the ice-cutter's bottom sprocket (Figure 1.23). The ice-
cutter comprises an unsinkable chassis, a driving winch with a rope,
a worm reduction gear with a steering wheel, an operating member with a
cutting chain, frame, augers, and a high-torque hydraulic prime mover.
Cutters are fitted on the augers. The steering wheel, via the worm

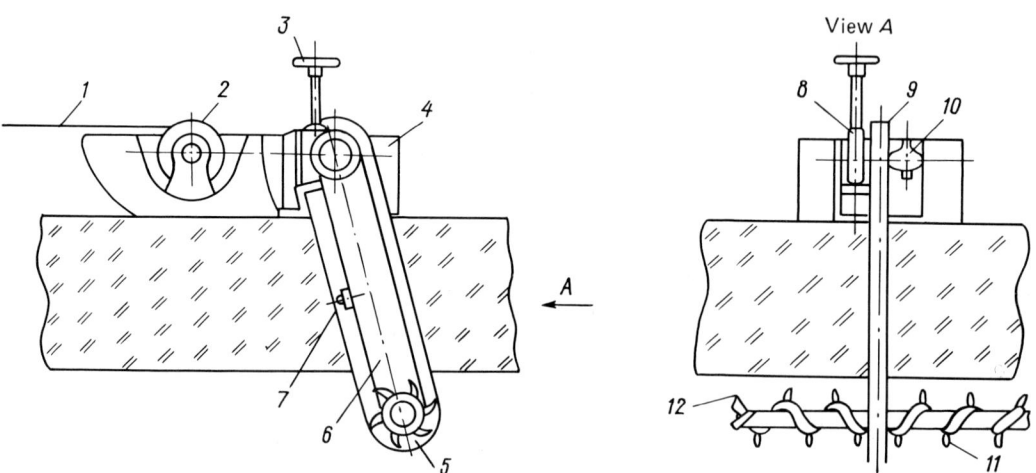

Fig. 1.23. Diagram of ice-cutting system [14].

1 - rope; 2 - driving winch; 3 - steering wheel; 4 - unsinkable chassis;
5, 11 - augers; 6 - frame; 7 - cutting chain; 8 - worm reduction gear;
9 - working member; 10 - hydraulic prime mover; 12 - cutter on augers.

reduction gear and sleeve coupling, turns and sinks the working member. The cutters on the augers facilitate cutting into the ice. The winch drive is engaged, and the moving system makes a cut in the ice.

Ice-cutting units based on the simultaneous operation of two side-milling cutters or two jibs, have been developed to cut long, wide lanes in an ice sheet. One of these units has been designed at the Research Institute of Water Transport [13]. The latter unit enables the cutting of a trapezoidal prism in the ice massif, breaking it into individual chunks, and removing the latter from the lane on to the ice sheet by means of a wedge-shear. The unit may be used to cut ice around vessels and ship convoys, as well as in freezing-out operations. The continuity of the ice-cutting and removal process ensures a considerable increase in performance as compared to other similar machines. The unit comprises a tractor with a mounted cylindrical bevel reduction gear and a jib hinged on the latter (Figure 1.24). The jib chain has cutters on the front and rear edges. The jib is shaped like a trapezoid with unequal non-parallel sides. The jib is perpendicular to the longitudinal axis of the tractor, and is connected to the latter by hydraulic jacks. A wedge displacer is hitched to the tractor. The shape of the end section of the wedge is similar to the shape of the channel cut by the jib.

When the jib is fixed in the initial working position, the unit

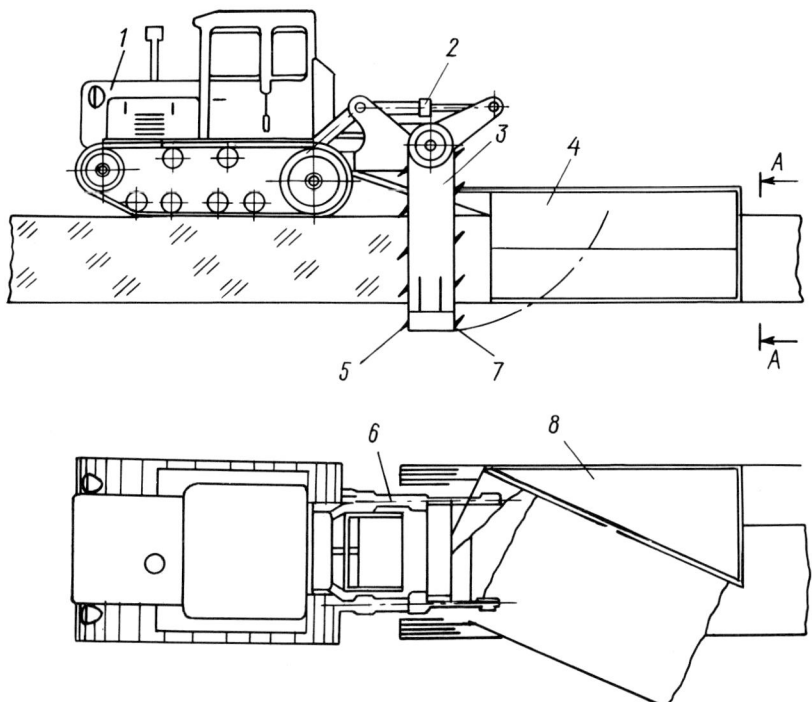

Fig. 1.24. Ice-cutting tractor-mounted jig machine D-75 [13].

1 - tractor; 2 - hydraulic jacks; 3 - jig chain; 4, 8 - wedge-displacer; 5, 7 - chain cutters; 6 - cylindrical bevel reduction gear.

begins to move along the route of the potential channel. The wedge-displacer breaks the forming cantilever ice bar into individual blocks, and displaces the latter by the skewed edge of the channel out onto the ice surface. According to calculation, the capacity of the unit may reach 300 m/h at an ice thickness of 1.5 m.

The experimental jigging unit Moroz (Frost) based on the trenching machine ETU-353, and the jigging unit BETN on the chassis of the wheel-mounted power shovel ETN-124, may be used for laying pipelines 70 cm in diameter under ice. These units are used to cut slots 14 cm wide in ice 1.0 m thick, and remove 1.5 × 1.0 m ice blocks. The cutting capacity is up to 130 m/h. The cutting speed is 1.5 m/s at a minimal power of 11-12.5 kW [2].

A unit designed to cut ice around convoys has been developed on the basis of a flexible cutting element made of a steel cable with round cutters (Figure 1.25). A round-link chain may be used instead of the cable.

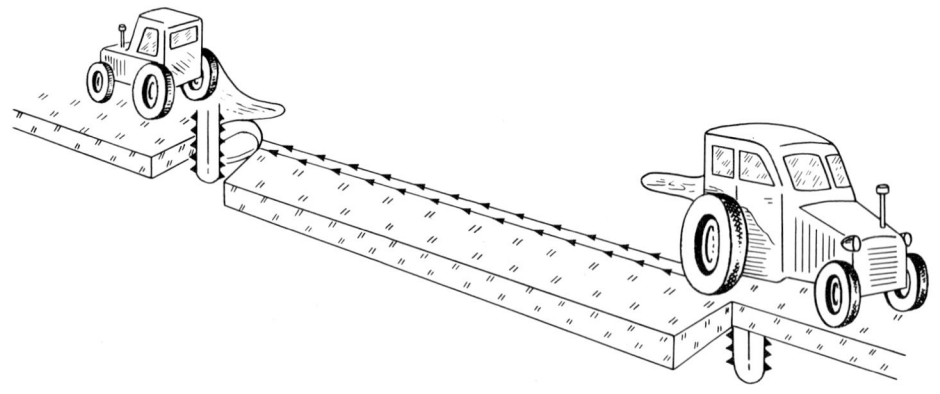

Fig. 1.25. Cutting ice with a flexible steel cable with round cutters [98].

The cutting element with a length up to 100 m is actuated by a power take-off. The width of the two cut slots is 2-5 cm each. The performance in cutting lanes is 5-6 times greater than that of units using jigging chains from mining machines, because the cutting process is realized simultaneously along the entire length of the flexible cutting element. After cutting, the remaining prism is pushed under the ice by the inclined surface of the moving wedge.

The flexible cutting element is the major part of tractor-mounted units for cutting ice around vessels (Figure 1.26). Comprehensive ice-cutting capacity (500-600 m/h) is attained mainly by cutting two narrow slots without breaking all the ice. The ice remaining between the slots is removed by a flexible band woven of steel cable, or a round-link

MECHANICAL DESTRUCTION OF ICE

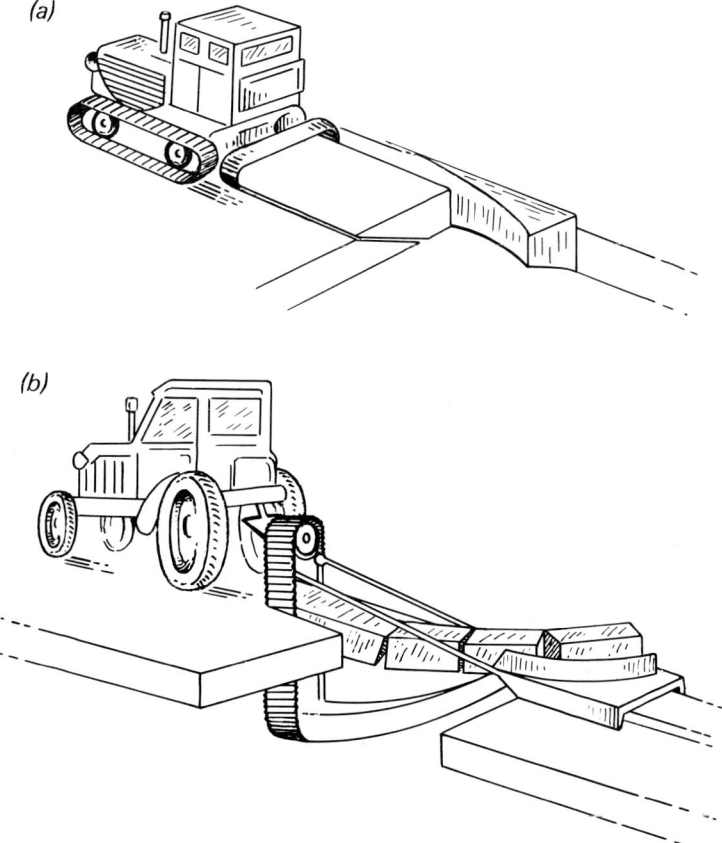

Fig. 1.26. Mounted units with flexible cutting element [98].

(a) - calibrated round-link chain with flexible cutters; (b) - flexible, woven steel cable band.

chain with round cutters. The figure demonstrates that a wedge removes the ice from the lane, and the broken-off ice chunk slides down an inclined surface, which requires less effort.

A system assembled directly in the dock has been developed to remove the ice cup from the underwater part of a vessel (Figure 1.27). When a vessel is docked, the ice is cut off by a continuous flexible cutting member (cable or chain) which is tensioned on pulleys by means of a block with a weight in the horizontal plane between the dock towers. The guide pulleys are arranged in such a manner that both branches of the flexible element are parallel in the horizontal plane. The guide pulley is connected to a winch, and the branches of the flexible element run in opposite directions, breaking and removing the ice cup while the vessel is dragged over the flexible cutting member.

The use of the double-jig machine BR-00-00, designed to cut ice 2-2.2 m thick, is of great interest [29]. The machine, consisting of

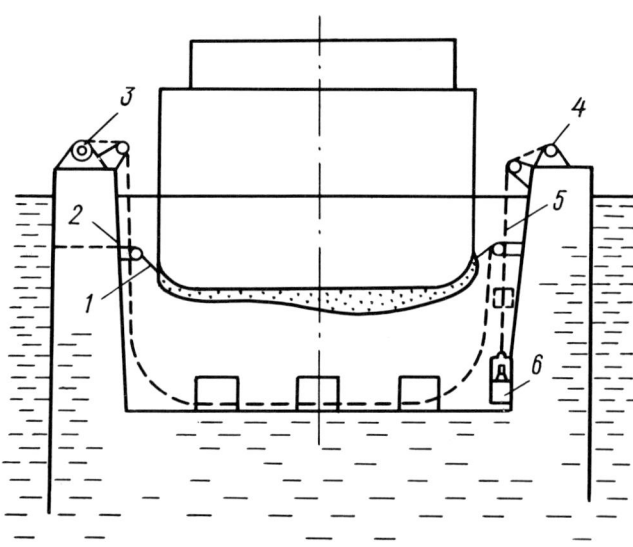

Fig. 1.27. System to remove frozen ice cup from vessel bottom [50].

1 - flexible cutting member; 2 - guide pulley; 3 - winch; 4 - tension mechanism; 5 - tension cable; 6 - weight.

T-100M tractor-mounted and built-in equipment (Figure 1.28), is designed to cut slots in frozen and solid soil, and to dig trenches and foundation pits.

SPECIFICATIONS OF RB-00-00

Machine base	T100M tractor
Working element	jig
Number of jigs	2
Total length of jigs (mm)	2800
Depth of slot cut in ice (mm)	up to 2200
Width of slot (mm)	140
Spacing between jigs (when working with two jigs) (mm)	700
Capacity (cutting ice 2.2-2.4 m thick) (m/h)	
maximum	90
average	50
Speed of working member chain (cutting speed) (m/s)	2.5
Dimensions (mm)	
length	7000
width (of tractor)	2460
mass (t)	16.4

MECHANICAL DESTRUCTION OF ICE   33

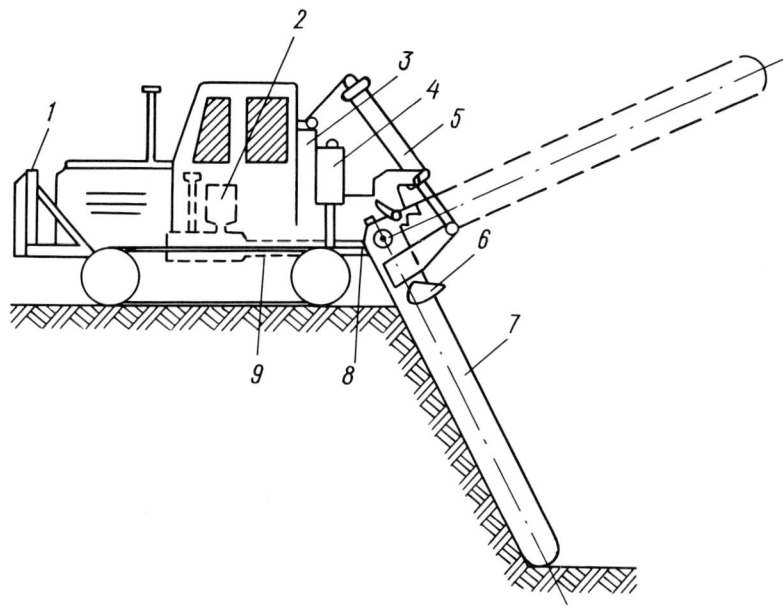

Fig. 1.28. Double-jig machine BR-00-00 [29].

1 - counterweight; 2 - hydromechanical reducing gear; 3 - hydraulic
equipment; 4 - oil tank; 5 - jig control mechanism; 6 - cleaning
device; 7 - jig; 8 - jig reducing gear; 9 - connecting shaft.

A slot 140 mm wide was cut in ice 2.2-2.4 m thick across a river
to lay a pull cable. The slot was cut by the jig machine, and the cable
was laid on the bottom simultaneously. The work was done as follows.
The pull cable was laid out on the ice 2 m away from the crossing. The
jig machine was set at the beginning of the crossing, which was marked
with stakes. A block was suspended on the lifted (inoperable) jig, and
the pull cable was fitted on it. When the jig machine moved, the pull
cable lowered under gravity into the slot and sank to the river bottom.
Two parallel narrow slots were formed by simultaneous operation of two
jigs. The ice between the two slots was cut by one jig into individual
chunks and extracted by the pipe-layer, thereby forming a lane.
  Experience in the operation of jig machines has made it possible
to set out recommendations ensuring safe operating conditions [29].
  Taking the overall mass of the machine into consideration, ice-
cutting operations should be carried out after testing the ice strength
in accordance with the 'Temporary Instructions on the Technology and
Organization of Building Underwater Crossings of Trunk Pipelines in
Winter' (All-Union Scientific-Research Institute for Construction of
Trunk Pipelines, 1968).
  The reduced ice thickness, as determined from the 'Temporary
Instructions', should be at least 60 cm in the case of one passage of a

jig machine to cut a slot 140 mm wide (one operating jig), or a slot 800 mm wide (two operating jigs).

In cutting wide lanes, requiring two or several successive passages of the jig machine over a weakened ice sheet, the permissible reduced ice thickness should be increased, depending on the ambient temperature. If the average air temperature in the previous three days was -5°C, the minimum reduced ice thickness should be increased to 1 m to permit operation of the jig machine for making wide cuts. As a rule, the ice thickness in northern regions of the USSR exceeds 1 m, so the use of a jig machine is most effective in these regions.

The jig machine should be at least 0.5 m away from the lane edge when cutting ice. The jigs should not remain in the slots when the working member chain is disengaged. The chain should be disengaged only when the jigs are removed. The pipe-layer should be at least 1 m away from the lane edge when lifting the ice.

1.2.2. Milling Methods

The journal <u>Inventor and Rationalizer</u> 1974 published the following remark on the work of the 'Ralsnemg' designing office at the A.A. Zhdanov Gorky Polytechnical Institute: 'The designing office, organized by A.F. Nikolayev, Dr. Sci., has developed and introduced original machines in the last 15 years that saved 80 million roubles.' Most of the machines designed for working in ice, snow and frozen soils have been developed on the basis of inventions made in this office, applying the principle of ice milling [105-109, 114, 118].

The machine LFM-GPI-I (Figure 1.29) has been designed on the basis of the GAZ-69 automobile. It is compact, and can break uneven ice in the form of ice reefs and hummocks with a height up to 1.5 m by means of mutually perpendicular spiral-milling cutters 0.5 m in diameter. Arrangement of the vertical cutter drive in the dead space rearward of the horizontal cutter ensures a working front 2 m wide. The mass of the machine is 1700 kg, the engine power 51.5 kW, the working speed in ice milling 90-250 m/h, the machine capacity (destruction of uneven ice and removal of ice aggregates from an aerodrome) 120 m^3/h and the specific power consumption 0:2 kW.h/m. The machine was approved for factory production after state tests at the SP-6 (North Pole-6) drifting research station.

LFM-GPI-34 is designed to break ice off vessels during winter mooring in backwaters and at ship repair facilities, when making crossings over frozen rivers, and while fishing in ice-covered water bodies. It ensures ice cutting, as well as snow removal from the work routes and fire emergency passages between vessels by means of a rotary screw unit.

The original design of the working member, which is a pin spiral-milling cutter with edges of a hard alloy, type VK-8, cuts a through slot at rotation of the milling cutters, advances the machine, and discharges the sludge under the ice, where the water stream carries it away.

Unsinkability of the machine is envisaged in order to ensure safe

MECHANICAL DESTRUCTION OF ICE 35

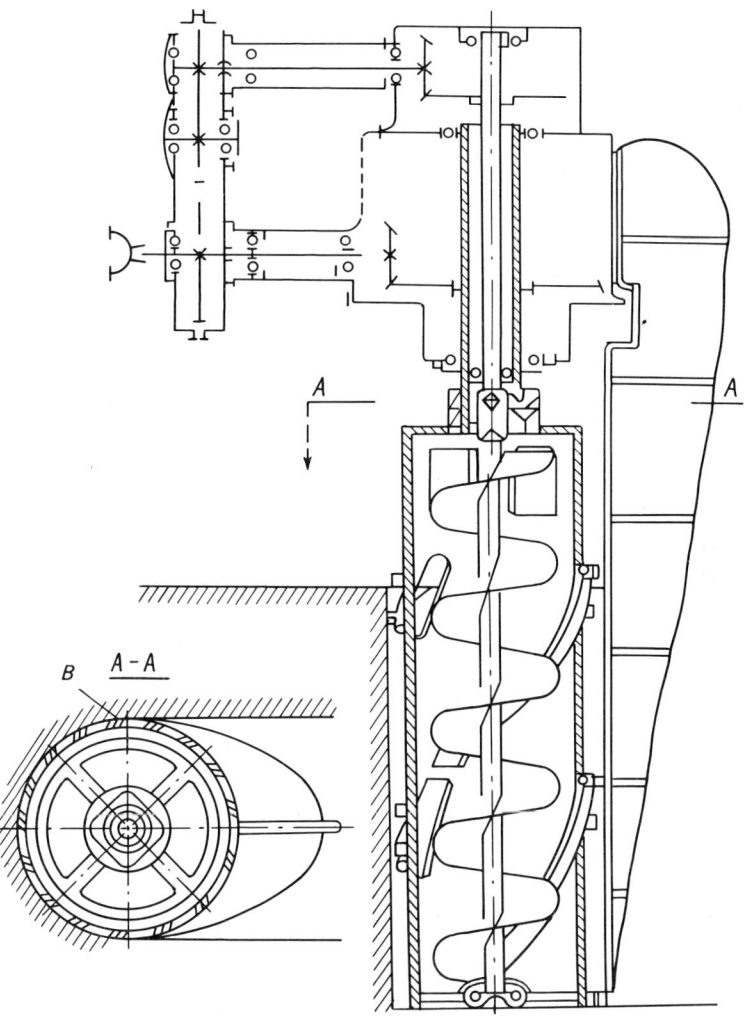

Fig. 1.29. LFM-GPI-I machine for trenching in ice [105].

operation. The body of the machine is hermetically sealed, and it is characterized by comprehensive displacement and stability in water.

SPECIFICATIONS OF LFM-GPI-34

Mass of machine (kg)	2450
Engine power (kW)	36.5
Cut ice thickness (m)	<1
Speed at cutting a through slot in ice (m/h)	
in first gear	125
in second gear	300
in third gear	470

Mill rotation speed (r.p.m.) 300
Specific power consumption in ice cutting (kW.h/m^3) 0.5
Operating width of rotary screw unit in snow clearing (m) 1.6
Capacity of rotary screw unit (in clearing snow 0.3 m deep) (m^3/h) 100

LFM-GPI-34 replaces 80-100 workers. The machine cuts a 1 m length 10-12 times cheaper and 80-100 times faster than manual labour.

The snow removal ice-milling machine LFM-GPI-41 is based on the GAZ-47 crawler conveyer (Author's Certificate 134,275 USSR).

The main feature of this invention is that pockets (holes) are made in the body of the mill near the cutter to eject the comminuted ice chips (without impeding the ice-cutting process), and an auger inside a pipe goes down under the ice. A machine with this mill can cut a trench 1 m deep, 0.4 m wide, and 300 m long in ice in one hour. The

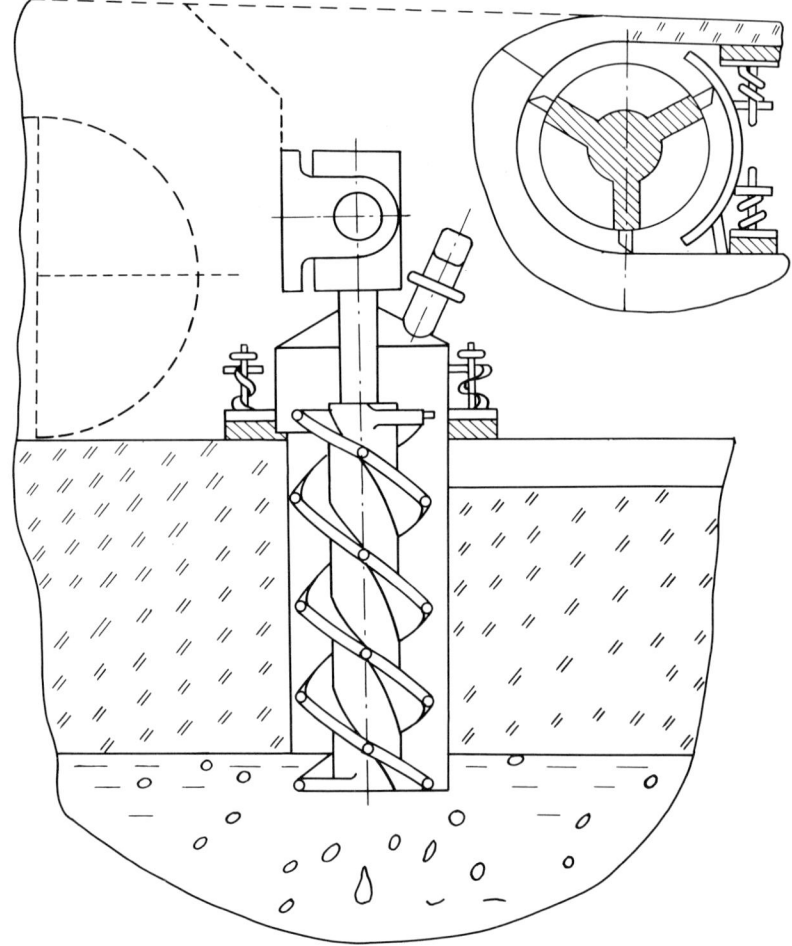

Fig. 1.30. Unit for cutting a channel in ice [105].

machine's mass is 4670 kg.

A unit for cutting a channel in ice (Author's Certificate 138,627 USSR) (Figure 1.30) comprises a self-propelled machine (automobile, tractor, etc.) with a driven auger in its front part to remove snow from the ice surface. The mill, which is suspended in the rear part of the unit, is turned from the horizontal to the vertical position. The distinguishing feature is that the mill is enclosed in a casing which is open from beneath and on the side of its cutting part in order to reduce the consumption of power in cutting a channel. A pipe at the top of the casing supplies compressed air to remove ice chips from the mill to the water. A flange with holes is welded to the uppermost part of the casing in order to seal it in the working position. Pins, freely inserted into the holes in the flange, are connected rigidly to a spring-loaded ring which is faced on the bottom side with rubber.

A system for cutting slots and trenches in ice (Author's Certificate 134,816 USSR) (Figure 1.31) comprises a disc with cutters around its circumference, and a frame with bearings for mounting and driving a mill. To increase the capacity and simultaneously reduce specific power consumption, the mill is hollow. Two discs are secured in a hub; a rim is secured on the discs, with a helical cylindrical gearing on the inside and two rows of cutters on the outside.

Equipment for trenching in ice (Author's Certificate 132,310 USSR) (Figure 1.32). A hollow tubular mill is used to cut ice. A cylindrical arrangement of a driving mechanism with a neck for the intake of air, and with a shutter on the inoperative side of the mill, is hinged on

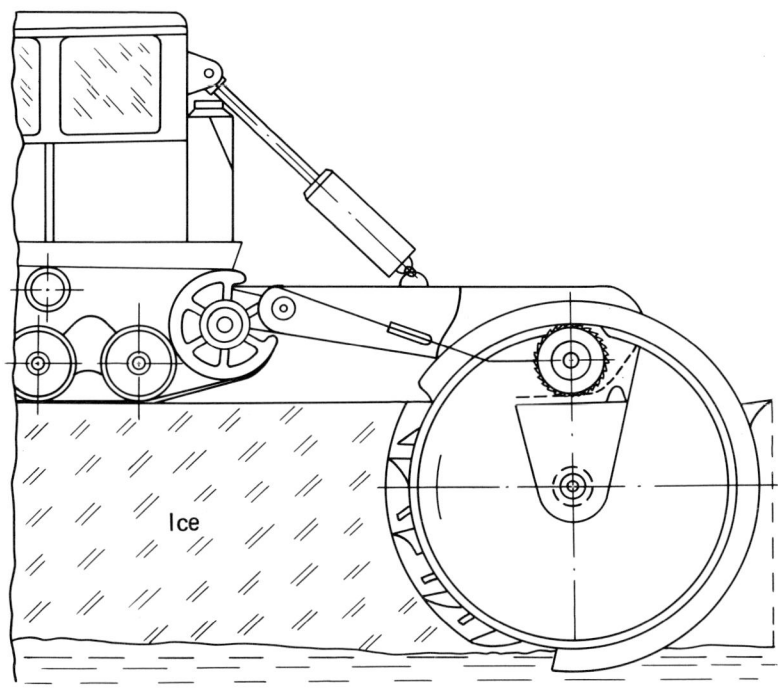

Fig. 1.31. System for cutting slots and trenches in ice [105].

the ice-cutting machine to remove ice chips and air. The mill body is connected to a hollow vertical shaft. Spiral - rakes are arranged on the outside of the shaft; cutters are fitted in the rakes as well as on the faces of the body. The guide blades of the air blower, arranged on the vertical surface, remove the ice chips from the trench.

The ice-cutting machine LFM-GPI-75 (Author's Certificate 327,289 USSR) is designed to break ice upstream from dams. The working part of the machine is a side-milling cutter, which cuts a slot 1.5 m deep and 0.15 m wide in ice. An 85 kW engine ensures a capacity up to 112.5 m^3/h. The machine's mass is 5.5 t.

The small-sized self-propelled ice-cutting apparatus SLU-80 (Author's Certificate 195,473 USSR) (Figure 1.33) is designed to cut ice around vessels, make lanes, and perform other ice-cutting operations. Two models of the apparatus have been developed and manufactured: one with an internal combustion engine, and one with an electric motor. The first model is self-contained and can operate without additional

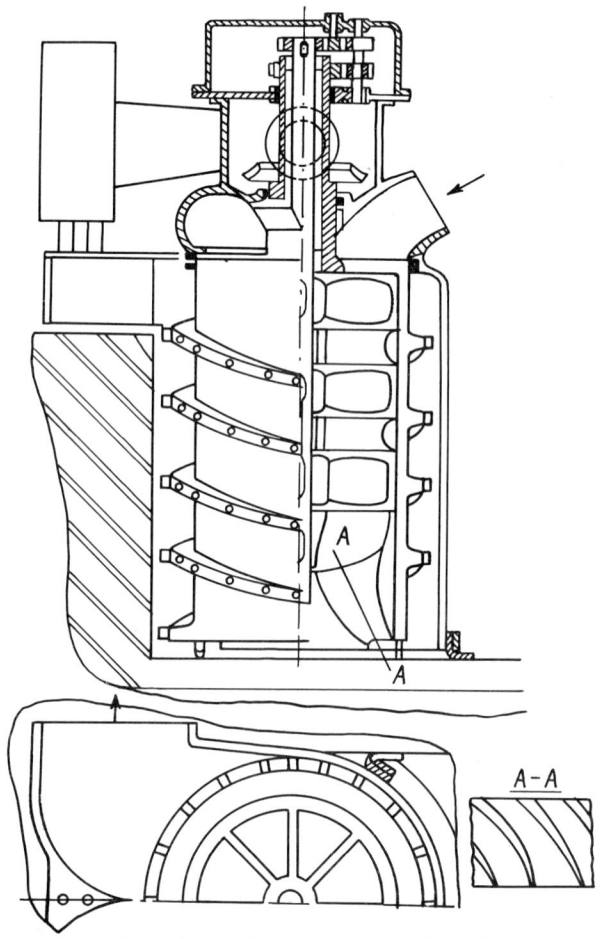

Fig. 1.32. Equipment for trenching in ice [105].

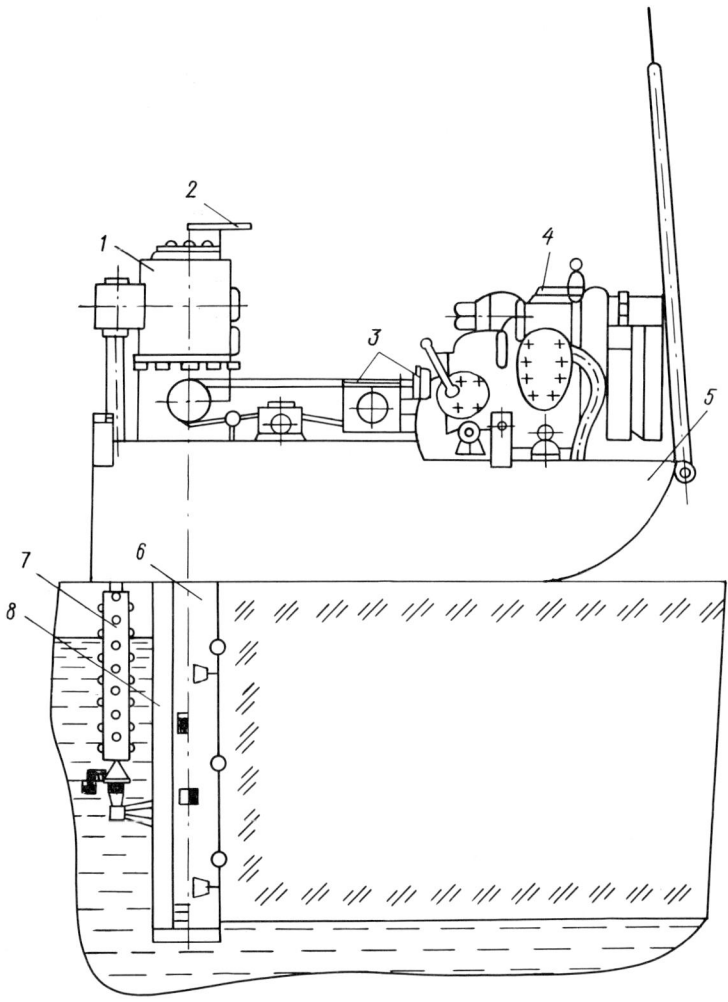

Fig. 1.33. Small-sized self-propelled ice-cutting apparatus SLU-80 [109].

1 - milling head; 2 - steering wheel of worm gearing; 3 - transmission; 4 - motocycle engine; 5 - hermetic sledge; 6 - end tubular mill with spiral conveyer; 7 - rollers with buses; 8 - casing.

fuelling for 10 hours. The apparatus is operated by one man. The SLU-80 apparatus is mounted on a hermetically sealed sledge, which is a welded frame of angle iron faced on the outside with sheet steel. A motocycle engine, (model M-72: stationary modification with forced air cooling) occupies the front part of the sledge. The engine torque is transmitted to the milling head, and the transmitted power is used in the milling head to power three components: the mill, the spiral conveyer, and the mover.

The working member is an end tubular mill with a built-in spiral conveyer to remove the sludge, entering the mill through windows ahead of the cutters. When the mill cuts into the ice, the sludge feeds up to the ice surface, but, when the mill cuts deep into the ice, the sludge feeds down under the ice. As a result, the water surface in the ice slot is clear.

Translational motion of the apparatus during ice cutting is ensured by a mover of original design. Two rollers, rotated by the same engine and fitted with sharp spikes, are arranged on the casing rearward of the mill. When the mill cuts into the ice, the rollers are turned so that they go freely into the cut slot. The mill cuts in to the end, and the rollers are turned by a special hand mechanism so that the spikes penetrate into the walls of the ice slot. Rolling over the walls without slipping, the rollers transmit translational motion via the casing to the entire apparatus. The characteristic feature of a mover of this type is that the flat walls in the cut slot make its bearing surfaces.

When the work is completed, the mill is removed to a stowed position by means of worm gearings, and the apparatus is ready for transportation.

The second variant of the small-sized self-propelled ice cutting apparatus SLU-80 is equipped with a 7 kW electric motor, which is connected to a three-phase circuit of 380 V, 50 Hz [109].

A mounted conical milling cutter (Figure 1.34) is used for trenching in ice [63].

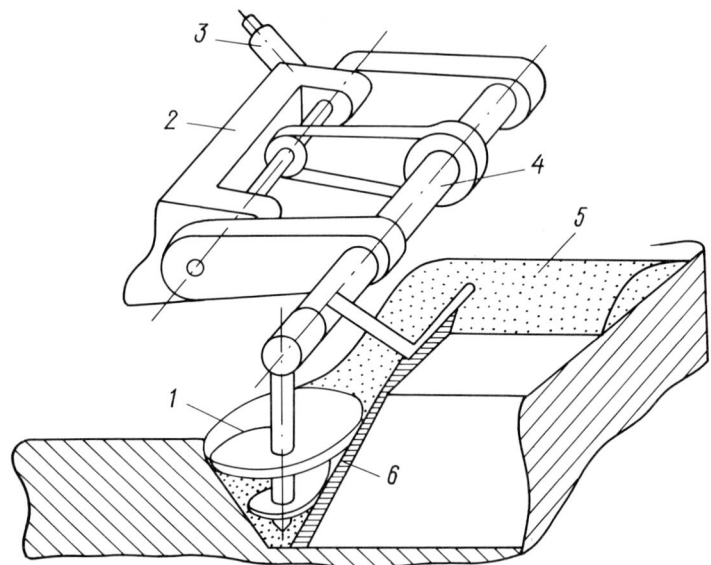

Fig. 1.34. Mounted apparatus with conical milling cutter for trenching in ice [63].

1 - conical milling cutter; 2 - tractor adapter, 3 -turning jack; 4 - mounted equipment; 5 - milling cutter casing.

Ice-milling machines may also be used for highway construction. The three-jig machine RRMZ, based on the T-130 tractor, is used to remove river ice when making embankments in winter that may be flooded in spring [152]. Ice with a thickness up to 2.2 m is cut at a speed up to 300 m/h.

The work of cutting an ice sheet is started at dates precluding the freezing of a lane more than 1/3 of its depth. For example, these dates are 1.5-2 months before the drifting of ice on the Northern Dvina River. The technological and economic advantages of this have been demonstrated by three years' experience of using ice-milling machines for dismembering the ice sheet with the aim of preventing ice-blocking at a water power development near a thermal hydroelectric station in Siberia, where 54 km of cuts have been made [118]. The optimal scheme of cuts was a network with longitudinal and transverse rows 200 m apart.

The advantages of ice-milling machines are as follows: ability to work on large areas; potentiality of breaking thick ice; reliability of operation irrespective of weather conditions; potentiality of working on shallow-water areas.

The disadvantages are the impossibility of creating large ice-free water surfaces; the impossibility of working on thin ice (less than 30 cm); the narrow slot cut by the machines freeze up rapidly, and the ice sheet recovers its integrity.

1.2.3. Marine ice milling

A brief description of marine ice-cutting methods, based on the study of patents and other information, is given in this section. Data on the use of this equipment are not available. The practical application of some devices seems doubtful, especially those designed by their inventors for use from an icebreaker navigating in ice (see for example, those described in pats. 3,768,428 and 3,913,511 USA). However, more comprehensive information on the results of these developments (including those that have not yet been sufficiently tested under practical conditions, or have not been tested at all, and are only protected by Author's Certificates and patents) will contribute to the emergence of new ideas about ice destruction in general.

A rotatable drum with hundreds of cutting knives 1.2-1.8 m long is proposed as a self-propelled floating drilling vessel. The drum, with a rotational speed of 10-15 r.p.m. (drive power 441 MW), is mounted on a stabilized column. The floating drilling vessel with this arrangement can navigate in ice 1.8 m thick at a speed of 5 knots, and keep station over the drilling location in the case of ice motion [80].

A sweep ice-cutter to cut a path through ice (pat. 3,768,428 USA) (Figure 1.35) is mounted on an icebreaker. When the vessel moves, the port and starboard ice-chipping cutters cut the boundaries of the channel, and the central cutter, traversing in a zigzag, cuts the internal ice field into sections that are easily cut into small floes and are moved apart by the hull of the icebreaker. The port and starboard cutters are made in the form of a set of side-milling cutters mounted on vertical shafts. The central cutter comprises two sets of

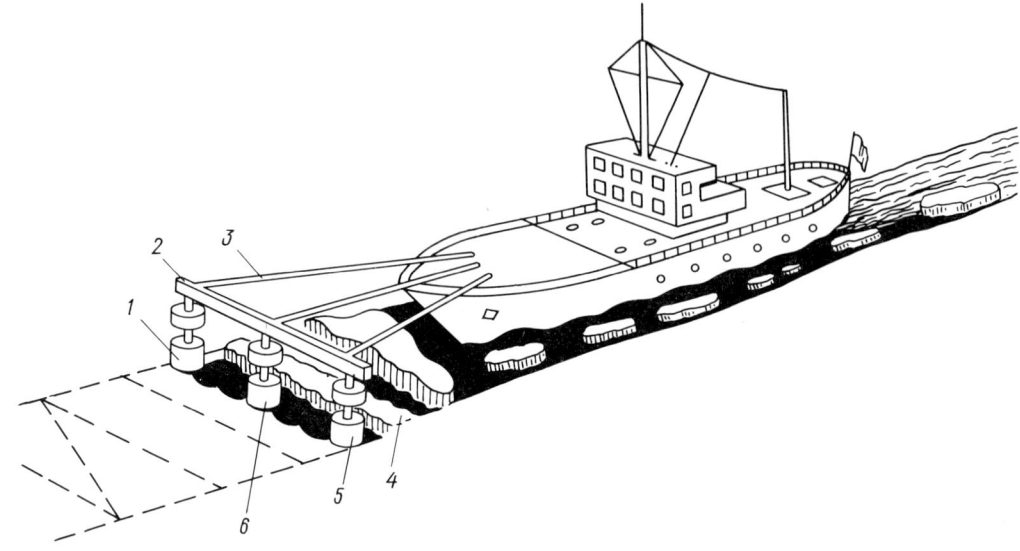

Fig. 1.35. Sweep ice-cutter to cut a path through ice [80].

1, 5 - port and starboard cutters; 2 - bearing frame cross-piece;
3, 4 - lateral struts of bearing frame; 6 - central cutter.

side-milling cutters rotating in opposite directions. The cutters are driven by electric motors.

Design methods for breaking an ice sheet by cutting frequently envisage members enabling the channel to be cleared of broken ice. The ice fragments may be transported in this case from the edge of the channel to an intact ice field.

A floating ice-cutting arrangement (pat. 3,913,511 USA) is a platform mounted on sealed drums which ensure buoyancy of the ice cutter. The drums are rotated by electric motors (Figure 1.36). Side-milling cutters with drives are mounted on the bow of the vessel. The ice is sawn into strips by milling cutters and the strips are then broken by the front drum. The fragments are picked up by a longitudinal inclined conveyer and are supplied to transverse conveyers that dump them on the ice field on both sides of the channel.

A model of an ice-cutting vessel, designed under the guidance of I.S. Peschansky [124, 125], was tested in the early 1960s in the experimental water body of the Arctic and Antarctic Research Institute. The vessel, designed for clearing a path through ice, was quite effective when operating in fixed (pack) ice of comparatively small thickness. The tests demonstrated in principle the potential and economic effectiveness of such a vessel operating at approaches to ports, in river mouths, etc.

An ice cutter, designed by American specialists (Figure 1.37), is a pontoon with a straight bow. Vertical-milling cutters are arranged on the bow to cut three longitudinal slots in the ice field during motion of the vessel. As a result, two cantilever ice beams are formed,

MECHANICAL DESTRUCTION OF ICE

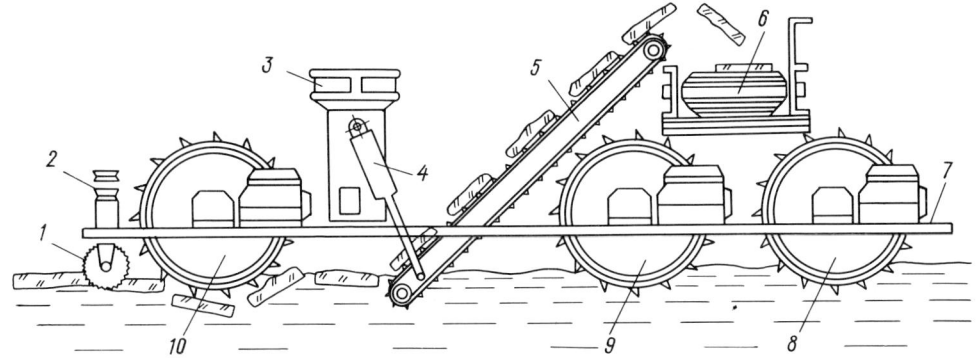

Fig. 1.36. Ice-cutter for clearing a path through ice [80].

1 - side milling cutters; 2 - drives; 3 - deck house; 4 - hydraulic cylinder; 5 - longitudinal inclined conveyer; 6 - transverse conveyers; 7 - platform; 8 - 10 - buoyancy drums.

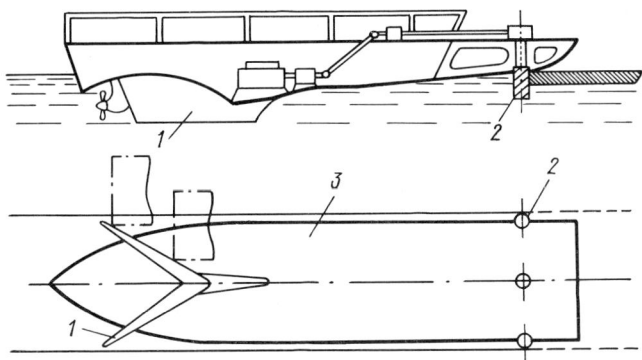

Fig. 1.37. Ice-cutter model with vertical-milling cutters [80].

1 - V-section keel; 2 - vertical-milling cutters; 3 - pontoon.

the width of each one being half the width of the channel. The beams bend downward under the effect of the inclined part of the vessel's bottom and break into rectangular slabs, which pass under the hull and are pushed aside under the intact ice sheet by the V-section keel. A path clear of ice remains astern of the vessel.

A ship for transporting freight under ice (pat. 3,768,427 USA) (Figure 1.38) comprises a submarine part for the freight compartment and engine room. The above-water part comprises a conning tower, connected to the submarine part by a mast which is also designed to break the ice sheet in order to make a narrow channel.

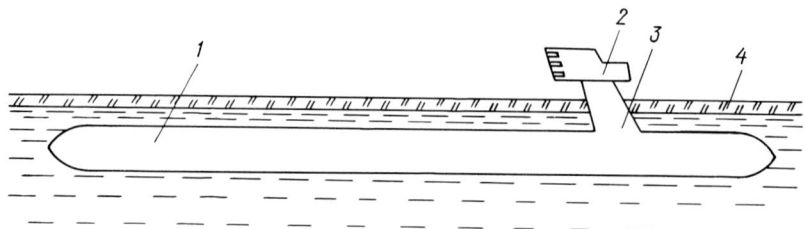

Fig. 1.38. Submarine ice-cutting tank ship [80].

1 - submarine cylindrical hull; 2 - conning tower; 3 - cutting mast; 4 - ice sheet.

An icebreaking tank ship (pat. 3,754,523 USA) is a semi-submerged catamaran. The ship's hull comprises a bow bulb, and the screw—rudder complex is in the stern. Both sides of the tank ship pass in the central part into inclined external sides, which in combination with the internal sides form two hulls that are interconnected by a superstructure. There is a free space between the superstructure and the ship's hull. The catamaran's hulls comprise an icebreaking bow. In clear water the tank ship navigates just like an ordinary vessel, with sides above the water surface. When navigating in ice, the tanks take in water ballast and the ship's hull submerges. The bow of the catamaran's hulls breaks the ice.

A marine ice-cutting system (pat. 4,005,666 USA) comprises multiple rotating cutter elements having many cutting edges spaced circumferentially around a common axis; the cutting edges engage the ice and dislodge the pieces. A means is available for directing fluid under pressure at the interface between each of the cutting edges of the cutter elements and the ice. The ejection of fluid results in ice cleavage.

In an improved model, (pat. 1,284,868 GB), multiple helicoidal screws, driven by a power plant, are arranged on the icebreaker's bow. The cutters of each screw are arranged along the edge of a blade. The helicoidal shape of the cutting screws ensures removal of ice from the upper or bottom part of the surface in the direction of icebreaker navigation.

Ice-cutting means for ships (Figure 1.39). A rod connected to a hoist—for example, a hydraulic cylinder—is deployed on a bracket fastened to a ship. A vane with a cutter is fitted to the bottom end of the rod. Level-detectors are fitted in the bottom part of the vane arrangement. The latter ensures cutter operation when the ship turns, and the detectors ensure the predetermined operating regime (predetermined cutting depth) of the ice-cutting device.

Cargo ship for navigation in ice (pat. 131,283 Norway) (Figure 1.40). The shape of the ship's hull ensures icebreaking from above and from beneath owing to the ship's variable draught. The spear-shaped stem breaks the ice from beneath at maximal draught WL_1, and

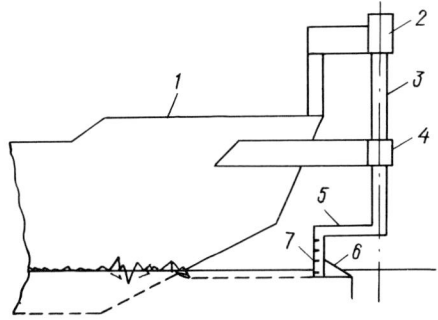

Fig. 1.39. Ship-mounted ice-cutting means [84].

1 - bow; 2 - hydraulic cylinder of hoisting mechanism; 3 - rod; 4 - bracket; 5 - vane arrangement; 6 - cutter; 7 - level detectors.

Fig. 1.40. Cargo ship for breaking ice by spear-shaped stem from above or beneath owing to variable draught [80].

WL_1 - maximal draught; WL_2 - minimal draught.

from above at minimal draught WL_2. The sides are distinguished by considerable flare in the regions of variable waterlines to enhance navigation in ice.

Icebreaker—ice-cutting tank ship (pat. 3,780,687 USA). The design is based on the principle of combining methods for breaking ice from above and beneath by ships with a variable draught.

The ship's hull (Figure 1.41) comprises a central cylindrical part, a conical bow, and a stern with a screw propeller and rudders. The ballast system is designed to create the required draught and trim for navigation in ice without cargo, when the bow surfaces and the ice breaks under the effect of its weight proper (Figure 1.41a), and for navigation with cargo, when the ship cuts the ice from beneath by the wedged edge of the bow (Figure 1.41b).

The principle of an ice plough has been used in the design of an icebreaker bow, ensuring ice cutting and clearing of a channel when the vessel goes astern (pat. 3,690,281 USA). The bow is share-shaped with concave sides above the waterline, running through the widest part of the stern frames below the load waterline (Figure 1.42).

The widened part of the bow is below the floating ice fragments.

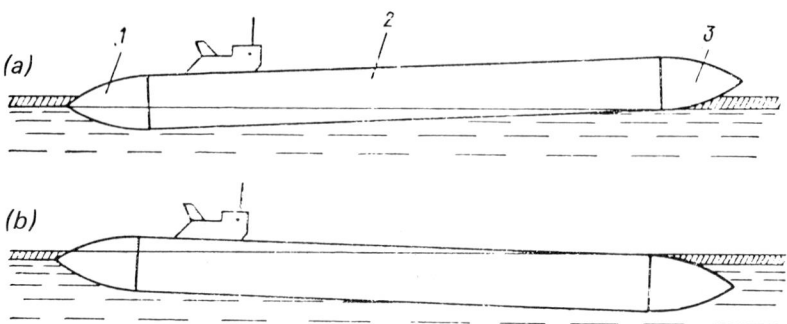

Fig. 1.41. Icebreaker-ice cutting tank ship with variable draught [80].
1 - stern; 2 - central cylindrical part; 3 - conical bow.

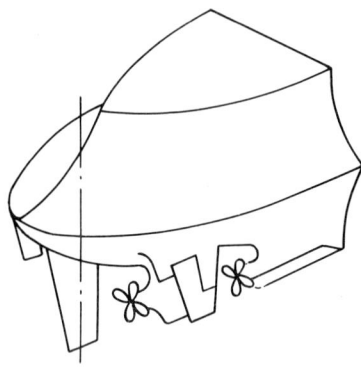

Fig. 1.42. Plough-share-shaped icebreaker bow [80].

When the vessel goes astern, its sharp edge moves the ice apart, safeguarding the screw propellers and rudders against damage. The concave sides ensure displacement of the broken ice to the edges of the channel. This design has also been patented in Finland.

The design of an icebreaker whose bow is below the ice sheet and is fitted with a device for cutting and transporting ice has been patented in the USSR (Author's Certiciate 147,468).

Their relatively low cost, and potential use on ordinary vessels, has encouraged the development of ice-cutting attachments.

An icebreaker—ice-cutting attachment, employing the effect of sliding on ice (Figure 1.43), is made in the form of a pontoon with skids on the bottom and a central knife-cutter. The attachment, pushed forward by the vessel, comes out of the water on the ice edge, owing to the small bow angle. When the attachment comes out on the ice, it presses down on it by its higher central skids. The attachment lowers as the central skids crush the ice and then the lateral skids, coming into contact with the ice, start to crush it, thereby widening the channel. When the ice under the skids is crushed the attachment continues its downward movement, crushing the ice with its sloping

MECHANICAL DESTRUCTION OF ICE

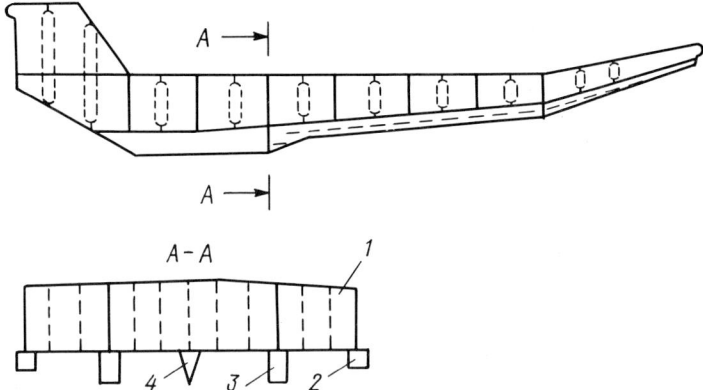

Fig. 1.43. Icebreaking attachment LPS [50].

1 - body; 2 - lateral skids; 3 - central skids; 4 - knife-cutter.

bottom. The central knife-cutter, which widens toward the stern, disintegrates the ice. The motion of the attachment moves the broken ice under the edge of the channel.

Tests have demonstrated that this attachment can make an absolutely clear channel up to 20 m wide in ice 70 cm thick, using one or two pushboats with a maximum total power of 1177 kW. The icebreaker attachment has certain advantages as compared to conventional icebreakers, the main ones being its low cost and simple design. The attachment can easily be manufactured at ship repair facilities. The pushboat without the attachment can also be used for its original purpose, and the attachment proper can be used as a pontoon.

Icebreakers with ice-breaking machines can cut through ice 2-3 times thicker that the unmodified icebreaker could cope with.

1.2.4. Chipping Techniques

Common hand dripping tools have been used for the destruction of ice under practical conditions since time immemorial: crowbars, axes, picks, etc.

The use of chipping techniques is based on the brittle failure property of ice under the effect of stress waves of sufficient intensity that may be induced in ice by a dynamic load concentrated in space and time, or a slowly applied static load, with an intensity sufficient for crack formation.

Irrespective of whether a force is applied to a fragile body slowly or rapidly, cracks propagate at a high speed when destruction commences. The maximal speed of load propagation through ice is the velocity of sound in that medium.

According to the theory, which has been confirmed experimentally [67], when at least one rupture occurs in a solid body subsequent crack formation is stimulated by the reserve of potential energy of elastic bonds that are always available in the body. The latter energy transforms into the surface energy of the fragments formed after disaggregation of the body. Owing to the phenomenon of reflection from the environment (water, air, or the metal body of the icebreaker), and interference of elastic waves in certain critical zones of bodies of finite dimensions, depending on their shape (ice field of comparatively small vertical size, or its fragments), the waves may focus, causing progressive increase of stress in these spots up to destruction. This raises the possibility of ice destruction by brief repetitive dynamic loads, as discussed in [17]. The recommendations there are based on the effect of removing the energy of elastic waves.

Theoretical and experimental results demonstrate that the displacement of particles in the environment, directed along a crack, conditions the formation of stress at the crack apex. An increase in the wave entrance angle reduces the component of longitudinal displacement, and the concentration of stress at the crack apex is correspondingly reduced.

Analysis of the character and specific features of ice destruction enables us to come to a conclusion on the existence of chipping surfaces (chipping craters) corresponding to the minimum energy of destruction for a certain type of cutter, physicomechanical properties of the ice, and characteristics of stress elastic waves [7, 72].

It follows from [32] that an increase in the initial chipping angle causes comprehensive reduction of the specific energy of destruction, but in practice this demands radical redesigning of the geometry of existing cutters and changes in the character of their interaction with the material. Two trends have emerged: removal of a suspension, comprising finely comminuted ice and water, from the zone of destruction by means of longitudinal grooves on the cutter, and utilization of an angle cutting tool, increasing the initial chipping angle by effective entrance of the lateral faces into the ice and shift of gently sloping craters to the surface. The recommended angle of the faces is 92-96°. The elastic wave from the left-hand face of the cutter is optimal for development of a crack along the right-hand face, and vice versa, which agrees well with experiment. The chipping propagates from the angle cutting tool into the crater without developing cracks in the ice massif, while a cutter with a triangular section develops numerous cracks in the ice, absorbing a comprehensive part of the energy of elastic waves. The latter waves will not participate in ice destruction as they do not meet a reflecting boundary near the cutter.

Hence, considerably less specific energy is consumed in ice destruction by the use of elastic stress waves than in the destruction of ice by cutting.

Unlike the machines discussed above, ice ploughs neither saw nor mill the ice, but chip it off by means of a wedge which is oriented horizontally and hitched like a plough to a floating tractor or a tractor-amphibian [171-173].

Laboratory and field tests of an ice plough have been carried out at the department of glaciology of the Institute of Geography of the USSR Academy of Sciences [171, 173]. The tests have demonstrated that

MECHANICAL DESTRUCTION OF ICE

the optimal cutting speed is less than 1 m/s (the traction force increases at a lower speed). The consumption of energy for ice destruction is about 0.03 kW/m^3, or 100 times less than in the case of ice-milling techniques.

The cutter is shaped like a truncated pyramid, widening in the direction of forward motion, and the working surface is formed by a plane at a cutting angle of approximately 25° and about a 3° inclination angle of the rearward part of the rod to the ice field. A device mounted rearward of the plough ejects snow—ice aggregates from the cut furrow. The width at the top of the triangular furrow is about 50 cm. The furrow's depth may be as much as 40 cm, depending on the ice thickness. The plough is used to make cuts in the top surface of an ice sheet, thereby reducing its strength (Figure 1.44).

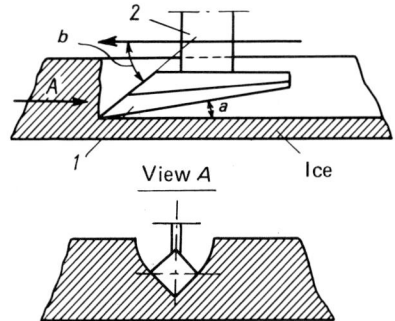

Fig. 1.44. Optimal shape of wedge for chipping thick ice [84].

1 - cutter; 2 - cutter handle; 3 - chipping angle.

Meltwater accumulates in narrow deep furrows, contributing to destruction of the ice sheet. The use of this device is most effective in preventing ice blocking in the case of floating ice on rivers, as well as in accelerating destruction of the ice sheet in the water area of ports and ship repair facilities.

Dismembering of a 10 km^2 ice sheet into blocks of 25×25 m [173] may be accomplished in five days with a two-shift working schedule. For example, the high capacity of an ice plough with a floating tractor GT-T ensures dismembering of the entire area of an ice sheet on a water body into 25×25 m blocks in 15 days. The capacity of vessels increased threefold in ports when the ice was weakened by ice ploughs [174].

Even at subzero temperatures, furrows cut by an ice plough melt through in 2-4 weeks and become channels. The process can be greatly accelerated if the operation of ice ploughs is combined with artificial intensification of melting by chemical and radiation methods.

In ship repair it is necessary to remove broken ice from the ship's bottom before docking. Mechanical means available for this include scrapers of different designs. The simplest is a metal beam, secured rigidly on the face of the dock between the towers so that the spacing between the scraper and the vessel's bottom at docking is about 100 mm. The use of this device is very complicated, because it requires continuous control of the spacing between the vessel's bottom and the scraper. A driver controls this spacing, and sends signals on the required change in the position of the pontoon deck in order to maintain a constant spacing. The operation is even more difficult during heeling or trimming of a vessel.

During comprehensive trimming, or in the docking of several vessels with different draught, the operating range of the scraper operation must be changed considerably depending on the spacing between the vessel's bottom and the scraper's axis. The hoisting mechanisms of the dock foundations must be modified to improve the operating characteristics of the scraper.

A surfacing ice-removing device (Author's Certificate 357,114, USSR), developed more recently can be used for vessels of widely differing draught. The device (Figure 1.45) comprises a series of ties, one end of which is articulated to the pontoon deck of the dock, and

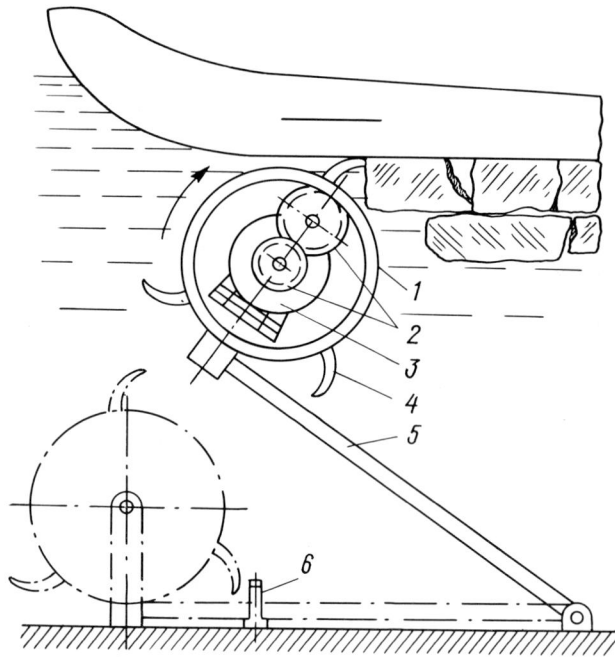

Fig. 1.45. Rotating scraper to chip ice from vessel's bottom [50].

1 - drum; 2 - gearing; 3 - electric motor; 4 - teeth; 5 - tie; 6 - stopper.

the other to the axle. A hermetic drum is mounted in the axle bearings, and an electric motor is mounted on a foundation inside the drum. The motor is connected via a transmission to the internal cylindrical surface of the drum. Toothed scrapers are fitted on the outside cylindrical surface of the drum, in a helical line. A stopper is fixed on the pontoon deck of the dock. When engaged with the tie, the stopper maintains the device in an inoperative position. The stopper and electric motor are controlled from the dock's tower. When the dock is lowered, the drum, having buoyancy in excess of its weight with the tie proper, strives to surface, but is held back by the stopper. When a vessel passes over the device, the stopper is let free and the device, turning on hinges on the pontoon deck, surfaces and presses itself to the vessel's bottom with a force equal to the drum's reserve of buoyancy. The electric motor is energized from the dock's tower and rotates the drum via a transmission so that the motion of the toothed scrapers under the vessel's bottom is directed in an opposite sense to the motion of the vessel proper. Owing to the dynamic effect the scrapers separate the ice floes from the bottom and, being arranged in a helical line, transport the floes to the vessel's sides. The floes surface, and the vessel is freely fixed on docking supports. The device may comprise two or more drums across the width of the dock. The drums may be differently oriented to the vessel, depending on the type of vessel under repair. The disadvantage of this scraper is the difficulty of manufacturing it, and especially of sealing its elements [50].

The capacity and effectiveness of some ice-cutting, milling and chipping techniques are shown in Table 1.3.

1.3. Drilling

The technology of mechanical rotary drilling is continuously being improved. Its capacity (depth of drill sinking per minute of continuous drilling) depends on ice hardness, drill rotational speed, axial load, type of cutting tool, and condition of the drilling tool's surface. The power consumed in drilling mechanisms during rotary drilling is expended on ice destruction, friction of the drilling tool against the ice, and pulling out the core.

Comprehensive experimental data have been accumulated and used to formulate certain notions of the theory of ice drilling. Complicated problems of designing cutting tools and machines for ice drilling have been solved, which is significant in prolonging the lifetime and reliability of these devices.

Much of the available experience is discussed in detail in [147-151], whence certain theoretical notions and specifications of ice-drilling units have been adopted.

The cutting tool for ice drilling is not unlike an ordinary drill, and the specifications generally adopted for twist drills may be used for ice-drilling bits. In the cutting edge of the bit it is necessary to distinguish the nose angle 2φ, the rake angle γ, the relief sharpening angle α, the cutting angle δ, and the wedge angle β. The nose angle at the point of a bit is usually 180°, or close to it. The rake angle, or the inclination angle of the drill's twist to the bit, may vary widely from zero to 60°. Investigations of bits with a predetermined angle have demonstrated that optimal results are ensured by a bit with an angle $\gamma = 40°$.

The main drilling factors are cutting speed V, bit feed S, chip

Table 1.3.
Effectiveness and capacity of some mechanical means for cutting, milling and chipping ice.

Ice destruction means	Power (kW)	Attending personnel	Speed of operation in ice of different thickness (m/h)			Furrow width (lane) (m)	Specific power consumption (kW h/m^3)	Capacity (m^3/h)	Result	Source of information
			0.5 m	1.0 m	1.5 m					
(1)	(2)	(3)	(4)	(5)	(6)	(7)	(8)	(9)	(10)	(11)
Manual cutting of channels	0.04	1	7	3	2	0.3	0.04	0.9–1.0	Through channel	[88]
Non-self-propelled ice-milling machine of Limend Works of Ministry of River Fleet	51.5	3	80	30	–	0.25	6.87	7.5	Through channel	[88]
Small self-propelled ice-cutting apparatus SLU-80	16.2	1	–	75	–	0.2	0.81	15	Through slot	[109]
Self-propelled ice-milling machine LFM-GPI-41 with with vertical knife and cam cutters	54.5	2	160	100	70	0.35	1.7	14–45	Through channel	[88, 105]

Ice destruction means	Power (kW)	Attending personnel	Speed of operation in ice of different thickness (m/h)			Furrow width (lane) (m)	Specific power consumption (kW h/m^3)	Capacity (m^3/h)	Result	Source of information
			0.5 m	1.0 m	1.5 m					
(1)	(2)	(3)	(4)	(5)	(6)	(7)	(8)	(9)	(10)	(11)
Ice plough with tractor-amphibian TP-90	66.2	2	4500	2000	–	0.6-1.5	0.03	1200-3400	Furrow not reaching bottom ice surface by 20 cm	[88, 171-174]
Ice cutting machine LFM-GPI-I	51.5		250	120	50	2.0	0.2	240-270	Even surface	[105]
Ice-cutting machine LFM-GPI-34	36.5			300		0.25	0.5	75	Through slot	[105]
Self-propelled ice-cutting machine LFM-66 with end-milling cutter	62.5			300		0.35	0.6	105	Through slot	[105, 190]
Ice plough with floating tractor GT-T	147	2	12000	5000	–	0.6-1.5	0.03	2206-6000	Furrow not reaching bottom ice surface by 20 cm	[171-174]

Ice destruction means	Power (kW)	Attending personnel	Speed of operation in ice of different thickness (m/h)			Furrow width (lane) (m)	Specific power consumption (mW h/m^3)	Capacity (m^3/h)	Result	Source of information
			0.5 m	1.0 m	1.5 m					
(1)	(2)	(3)	(4)	(5)	(6)	(7)	(8)	(9)	(10)	(11)
Self-propelled ice-cutting machine LFM-RVD-GPI-72 with tubular mill	85	2	—	400	—	0.36	0.6	140	Through slot	[105–109]
Self-propelled ice-cutting machine LFM-GPI-75 with side-milling cutter	85	2	—	—	500	0.15	0.76	112.5	Through slot	[105, 190]
Non-self-propelled milling machine	22	1	—	75	—	0.20	0.47	15	Through slot	[105, 190]
Jigging unit Moroz based on trenching machine ETU-353, and jigging unit BETN based on wheel-mounted power shovel ETN-124	12.5	2	—	130	—	0.14	0.7	18	Through slot	[2]

area f, drilling depth H, feed force P, torque M, and power consumed in drilling N.

The cutting speed is the displacement of the cutter's cutting edge in respect to the treated surface per unit of time ($\underline{V}$ m/min). In the case of drilling ice, the cutting speed is the speed at the periphery of the bit, i.e.

$$V = \pi Dn/1000,$$

where V is the speed (m/min); D is the bit diameter (mm) and $\underline{n}$ is the bit rotaitonal speed. Experiments demonstrate that the axial force and torque do not change significantly with different rotational speeds of the bit at the same feed. The bit feed is the distance the bit passes per revolution (S mm/rev), or per unit of time (S mm/min).

A bit may be fed manually or by a mechanical drive. Because the feed at the given power of the ice bit depends on the geometry of the cutting tool and mechanical properties of the ice, it is very important to be able to vary the feed so that the ice bit can work continuously in an optimal regime. Hence, it is recommended that several feed speeds should be provided in ice-drilling units.

The chip area f is the feed function. The less the feed, the less the chip section area, and vice versa.

The chip section area is

$$f = SD/2,$$

where S is the feed and D is the bit diameter.

When the bit sinks into the ice, the chips are ejected from the hole less forcefully, and the feed force, torque and power increase. As a result, various chip spreaders, augers, and other devices have been proposed in the design of ice drills, contributing to the ejection of chips from the hole. This problem has not been completely solved, which is the reason for certain specific features of the drilling process.

The chips formed are completely ejected from the hole at the beginning of the drilling process, and the passages in the bit proper remain free, causing no obstacle to drilling. When the bit sinks into the ice, the rate of chip ejection is reduced and eventually chips are not ejected at all. The depth at which chips cease to be ejected depends on several factors: bit rotational speed and chip size, ice conditions, bit design, etc. The zone in which the chips are ejected totally is called the zone of free dilling, and the zone in which chip ejection is impeded, and then halted totally, is the zone of constrained drilling. It is difficult to draw a clear border between the two zones, but, it may nevertheless be stated that the dynamics of the process differs between zones.

The effect of the drilling depth on the change of power has been investigated by drilling freshwater ice in a water body at -27°C with a bit 250 mm in diameter (Figures 1.46 and 1.47). The ejection is weak, while the power required increases rapidly. The increase in axial force, torque and power is explained by worsening chip ejection from the hole,

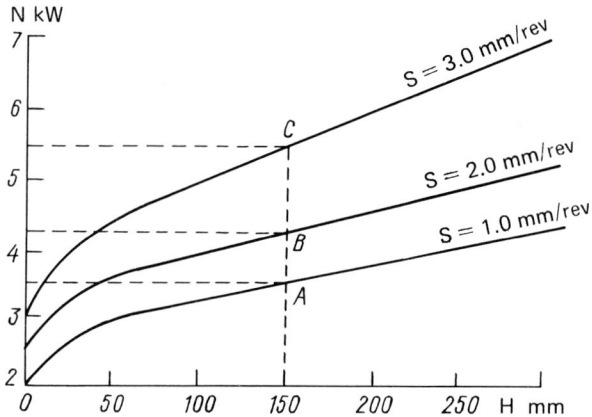

Fig. 1.46. Dependence of drive power (N kW) on drilling depth (H mm) and feed (S mm/rev) of bit [148].

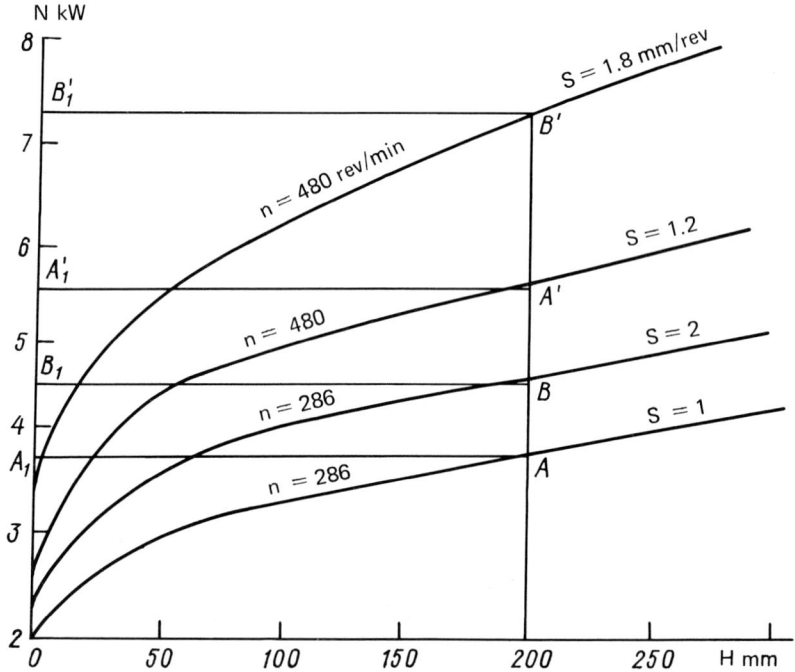

Fig. 1.47. Dependence of drive power (N kW) on drilling depth (H mm), feed (S mm/rev), and rotational speed (r) of bit [184].

which impedes separation of newly formed chips from the bit cutters. The chips are compacted and the hole is increasingly blocked up while the bit sinks into the ice.

The drilling process in the zone of constrained drilling depends greatly on the power reserve of the ice-drilling plant. If the plant's power is great, the chips may be compacted to such a degree that they turn into a solid ice mass, and, if the power reserve is exhausted (before the drilling of a passage is completed), the bit stops rotating and breaks down. Cases are known when it has been necessary to break the ice around the bit by means of an ice pick in order to pull it out of the hole.

It is necessary to use a method of drilling in stages until the problem of removing the chips from the hole is finally solved.

This method involves drilling a hole in several passes. The drill is pulled out periodically, simultaneously removing the chips that have accumulated in the hole. Naturally, the greater the number of stages, the lower the power characteristics of the process, but on the other hand additional time is necessary for pulling out and sinking the drill. It is therefore necessary to choose an optimal solution, i.e. it is necessary to work out a system of staged drilling for each ice drill, depending on its mass, drive power, and operating conditions. The ice drills are pulled out and the chips are removed in some cases every 200-300 mm; with more powerful drills, every 400-500 mm is sufficient.

The nomograph (Figure 1.48) makes it possible to determine the torque, feed, and axial force by the given drive power of the ice drill, spindle rotational speed and ice temperature. We mark the unit's capacity on the bottom scale, for example, 5 kW (point A). A vertical line is drawn from point A to intersect with the sloping line indicating 600 r.p.m. (point B). A horizontal line is drawn left from point B until it intersects with the inclined lines indicating ice temperature. Assume that drilling is performed at an ice temperature of -10°C (point C). We produce the horizontal line to the left from this point until it intersects with the ordinate axis, indicating the value of the torque (point D). This point corresponds to M_t = 8 kN.cm. Drop a vertical line from point C to intersect with the abscissa, indicating the feed value (point E). This point corresponds to a feed value of 3 mm/rev.

Having determined the feed value, draw a vertical line upwards from this point to intersect with the inclined dashed line t = -10°C (point F). A horizontal line is drawn from point F to the left to intersect with ordinate axis, indicating the value of axial force (point G). This point corresponds to P = 2.3 kN. Thus, the values of M_t, S and P have been found from the data on N, n and t.

The nomograph has been drawn for a drill 250 mm in diameter, but similar nomographs may be drawn for drills of any diameter. It should be borne in mind that this is the first attempt to systematize such experimental data and present them in a conveniently usable form.

The geomtery of the tool, removal of chips, drilling depth, etc. have not been fully considered in plotting this nomograph. The accumulation of experimental data will enable correction of the nomo-

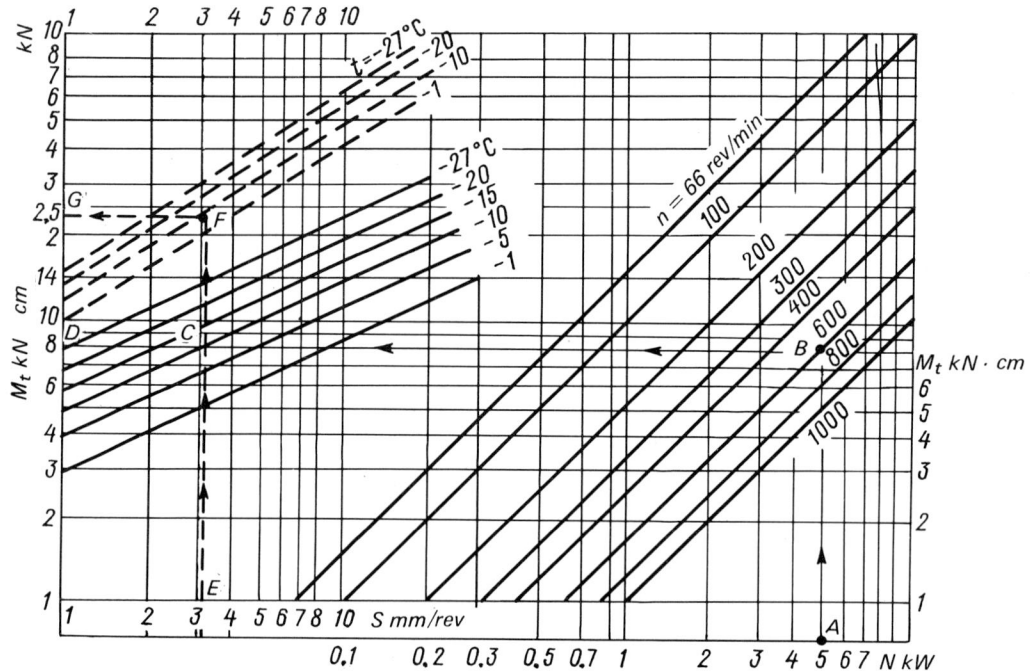

Fig. 1.48. Nomograph for selecting parameters of drilling process: torque (M_t kN.cm); feed (S mm/rev), and axial force (P kN) for given values of drive power of ice drill (N kW); spindle rotational speed (n r.p.m.), and ice temperature (t°C) [148].

graph and a more intelligent approach to the drilling of ice [148].

The consumption of power for drilling a hole may be considered to be proportional to the volume of broken ice. In order to reduce power consumption, holes should be of minimal necessary diameter. New methods of drilling holes in ice should also be developed.

In particular, attention should be paid to the potential advantages of drilling ice by means of a circular milling cutter. If the drilling depth is H and the hole diameter D, then the volume of ice turned into chips is $\pi D^2 H/4$. But there is no technological necessity to turn the entire ice volume into chips; this is only a consequence of utilizing a simplified drill.

When utilizing a circular milling cutter, a plug is cut out of the ice sheet and only a small volume of ice is turned into chips:

$$(\pi D^2/4 - \pi d^2/4) H,$$

where D is the cutter outside diameter and d the inside diameter.

The difference in the ice volume turned into chips is therefore

$(D^2 - d^2)/D^2$. About nine times less power is consumed in drilling a hole by means of a circular milling cutter at D = 350 mm, d = 330 mm.

V.I. Fedotov [160] distinguishes the following types of ice drills, depending on their use and operating conditions: twist drills; jumper or hollow drills; core drills; hand drills of other designs; mechanical drills with a drive proper.

Small holes (with a diameter up to 50 mm) are usually drilled in ice by means of twist drills, which are most convenient because of the continuous removal of the chips to the surface and a high drilling speed. Different types of ice drills, enabling the boring of only shallow (<1m) holes with a diameter of 120-150 mm, are used rarely. Twist drills have not justified themsleves in operation because they turn all the ice in the hole into chips, which requires much power, especially in drilling large holes.

The ice-drilling device GGI comprises a brace, a steel rod 12 mm in diameter, marked in centimetres and a twist drill 25 mm in diameter. The time required to drill through ice 0.5-0.6 m thick is 2-3 minutes.

The ice GGI drill comprises a bit 100 mm in diameter and a metal handle at the top of the drill. The procedure is to hit the drill head with a sledge hammer and then turn it, as in the use of a jumper drill.

Ice drill GR-7. The twist bit of the drill is a long double-threaded screw with a 200 mm lead and 63-64 mm in diameter. Rotation is accomplished by means of a brace which has a chuck to clamp the bit. The screw's concave-sharpened thread, of triangular section, forms two cutting edges with a 130° central angle. The cutting edges are interrupted in the centre by a rectangular cut. The bottom planes of the cutting faces are sharpened at a 20° angle to the horizontal plane. The diameter of the bit point is 69 mm at a distance of 20-30 mm from the cutting edges, in order to avoid excessive friction against the hole walls during drilling.

SPECIFICATIONS OF GR-7

Diameter of drilled hole (mm)	68-70
Drilling depth (cm) (with additional bit)	12-180
Drill mass (kg)	
with 1200 mm dia. bit	6
with 1800 mm dia. bit	8.5
Overall dimensions (mm)	
with 1200 mm dia. bit	70×90×1700
with 1800 mm dia. bit	70×290×2300

Ice drill GGI-47 (Figure 1.49). The bit is manufactured of strip steel. It is 105 mm in length and is designed for drilling ice 1 m thick. The diameter of the cutting faces exceeds the diameter of the upper part by 2-3 mm. The bottom 25-30 cm of the bit is hardened. The cutting faces of the bit form an angle of 140°. The bit comprises a threaded point with a nut at the top. The upper middle arms of the brace are fitted with wooden handles at the bottom of the chuck for coupling the brace to the bit. A washer-gauge, designed for checking correct

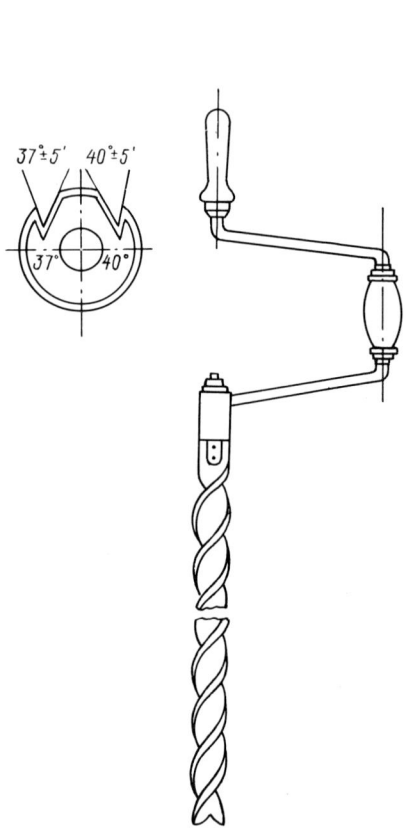

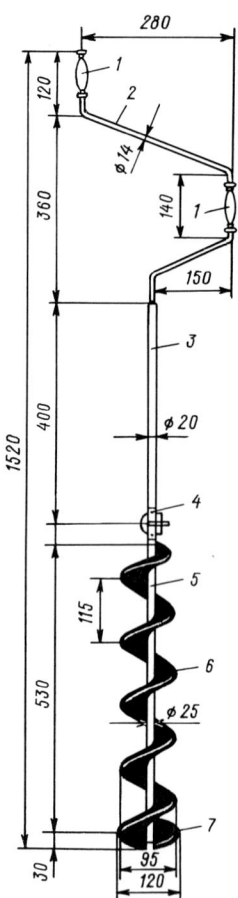

Fig. 1.49. Ice drill GGI-47 [160].

Fig. 1.50. Ice drill with auger [146].

1 - rotating hollow plastic handles;
2 - brace tommy bar; 3 - pipe;
4 - locking device, 5 - coupling rod;
6 - auger; 7 - knife.

sharpening of the bit cutting edge, is fixed with a nut at the point where the bit is coupled to the brace during assembly. The ice drilling speed is about 0.3-0.4 m/min. The hole diameter is 42-45 mm. The overall dimensions of the drill are 70×254×1460 mm.

Ice drill GU. The drill's diameter is 40-60 mm. An iron pipe rod 25 mm in diameter is welded to the 1000 mm long blade. The drill is fitted with 1500 mm long extension rod sections. The drill is rotated by a tommy bar, or Kazantsev's device. The blade and the cutting part are similar to the respective parts of the GGI-47 drill.

MECHANICAL DESTRUCTION OF ICE

Hand ice drill, model 'Toros' ('Hummock'). The diameter of the drill is 75 mm. The blade is 800-1500 mm long. The drill is rotated by a brace.

Marine ice drill of the Arctic and Antarctic Research Institute, designed by Sokolnikov and Multanovsky. The drill's diameter is 50 mm. The blade is 500-1500 mm long. The total length with the rods is 5-6 mm. It is rotated by a brace.

Ice drill with an auger (Figure 1.50). This is manufactured by Rosokhotrybolovsoyuz enterprises. It can be folded in two for transport [146]. The ice drill comprises two basic parts: a brace and a bit (an auger with a cutting part). The bit consists of a coupling rod, which is a thin-walled steel pipe 25 mm in diameter. A 530 mm long auger with a 115 mm pitch is welded around the rod. The auger is made of 1.5×35 mm section steel strip. The ice drill is used to drill holes of 120 mm diameter.

The auger and coupling rod have a knife at the end. The knife blade is interrupted in the middle by a 10 mm long cut, forming two cutting edges.

The knife is not detachable — it is welded to the coupling rod and the auger. It is manufactured of 3 mm thick steel and the design of the knife enables it to be sharpened. The chamfer sharpening angle is 20°. The size of each of the two cutting edges of the knife is 10×60 mm. The cutting edges are arranged at a radius of about 60 mm and the central angle between them in the vertical plane is about 230°.

SPECIFICATIONS OF ICE DRILL WITH AUGER

Maximal drilling depth (mm)	1050
Hole diameter (mm)	120
Bit rotation	
right-hand	clockwise
planetary	revolution
Mass (kg)	2.7
Overall dimensions (mm)	
in working (assembled) position	1520×320×120
in transport (field transit position with plastic tip)	890×320×130

An ice drill with an auger (as compared to an ice pick) decreases the labour involved by approximately four times, and has certain advantages over other ice drills. A 1 m deep hole is drilled in about 5 min. The specific feature of the ice drill is that, owing to the auger, when holes are drilled in dry ice even when the thickness is in excess of 500 mm (exceeding the length of the drill), the comminuted ice is ejected to the ice surface and it is unnecessary to clean out the hole. A hole 1 m deep is drilled in dry ice in one pass without seizure of the drill in the ice.

It is hard to drill wet ice with the ice drill, so in this case a multistage method of drilling is used. The ice thickness is not drilled through immediately, but in several passes, depending on the ice thickness, density, etc. The ice drill is pulled out of the hole after every 20 cm to clean it of comminuted ice.

Drilling with an auger ice drill is facilitated by having two handles so that the brace can be rotated by both hands. Also, if the upper arm of the brace's tommy bar is extended to 280 mm, that also facilitates drilling (the SPL-I ice drill, for example, has a 200 mm arm). The upper arm is displaced by 130 mm in relation to the ice drill axis, so it is impossible for two workers to drill one hole together.

Before starting operation, the ice drill is assembled in the working position and fixed with a nut and a handle. The knife should be sharp, without any nicks, burrs, or other defects which might impede the drilling process. The snow should be cleared away if it is more than 20 cm deep. The ice drill should be rotated uniformly, without jerks, at a speed of 1-1.5 rev/s.

It is necessary to be careful when finishing the drilling of a hole, when the knife approaches the ice bottom surface, so as to avoid breakdown of the ice drill. The drill should be rotated at a slower speed and less vigourously. At the end of the drilling operation, when the knife is at the bottom surface of the ice sheet, the ice drill should be pushed down with some reciprocating motion to wash out the hole. The ice drill should be periodically cleaned of frozen-on ice by light impacts on its bottom (working) part.

Finnish researchers have used twist drills with an unusual shape of the cutting part, having a depression instead of a projection.

Usually two drills were used for sounding: one 2.5 cm in diameter, and one 5.5 cm in diameter. The cutting edge of the small drill started in the middle of the cutting part (bit point), while the cutting part of the large model had a slightly inclined groove in the middle (Figure 1.51). The drilling speed with this drill was quite high.

Twist ice drill (Figure 1.52). This has been designed at the US Laboratory of Scientific and Engineering Investigation of Arctic Regions. The drill, which has a diameter of 25.4 mm, cuts easily into sea and glacier ice at 0.5 r.p.m. The drill's twist is cut, instead of being made of strip steel. The outside diameter and the cut groove are ground. The bit's blade is made of stainless steel. The drill is fitted with extension rods and is rotated by a brace. Methyl alcohol is occasionally poured into the hole during the drilling process to reduce freezing of the drill to the walls of the hole.

Ice drill with auger. An ice-drilling auger has been patented in the USA (pat. 3,929,196 USA), a manual auger drilling machine (pat. 4,057,114 USA), and an ice drill with adjustable cutting edges (pat. 978,180 Canada).

Jumper drills, or hollow drills, as they are called by some ice researchers, are hollow pipes whose rear end comprises slightly set sawlike teeth.

American hollow pipe drill (Figure 1.53). This is a jumper drill in design. It is used for drilling firm and melting sea ice, but it can also drill solid ice at a low temperature.

A view of this ice drill with extension rods, in a slightly modernized form, is shown in Figure 1.54. The ice drill's body is made of a 508 mm long steel pipe. The minimal diameter, permitting ice to be ejected from the drill's cavity, is 44.5 mm. Twelve teeth are cut

MECHANICAL DESTRUCTION OF ICE 63

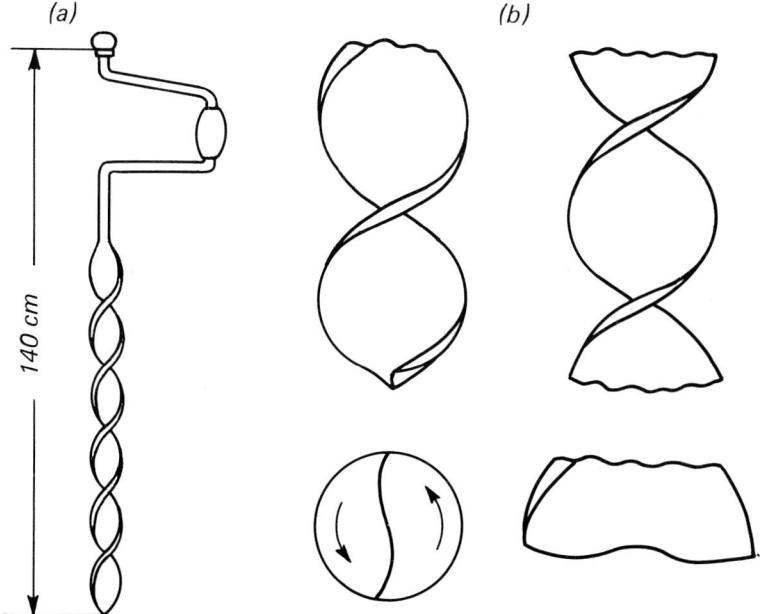

Fig. 1.51. Finnish ice drill [160].

(a) - general view; (b) - cutting part of drill.

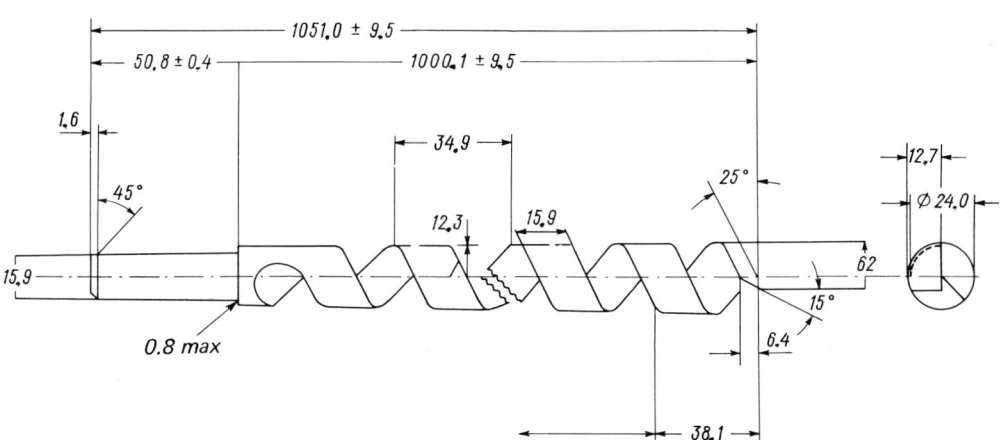

Fig. 1.52. American twist ice drill (dimensions in mm) [160].

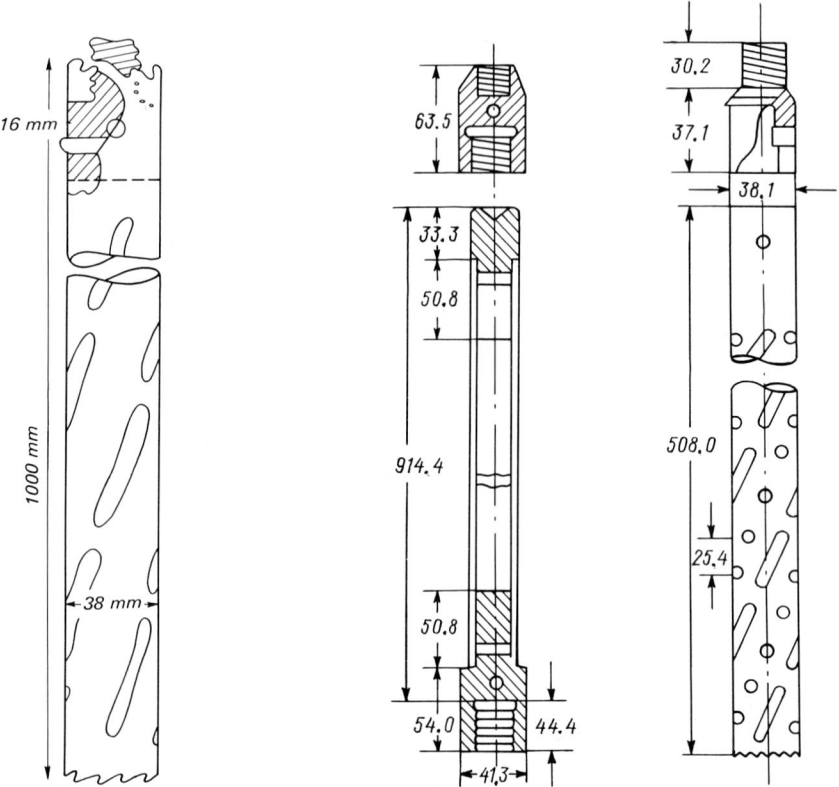

Fig. 1.53. Hollow pipe drill with longitudinal orifices [160].

Fig. 1.54. Hollow pipe drill with longitudinal and round orifices complete with extension rod [160].

and set by hand. The teeth can work in ice for a long time without additional sharpening. The drill's body comprises several orifices (d = 11.1 mm) and slots to ensure free space to remove the chipped ice from the hole. The brace and rods are coupled to the body of the drill by means of a screw thread.

The work is impeded at an ice density in excess of 800 kg/m^3, and great effort and wide practical experience are required when drilling still denser ice. The maximal drilling depth with 20 extension rods is 18 m. It is therefore more effective to use two operators to drill dense ice to a depth of 45 m. A powerful hoist is necessary in the case of deeper drilling. The total mass of the drilling equipment required to drill to a depth of 18 m is 36 kg, and the mass of a 1 m duralumin extension pipe is 1 kg.

A core drill is a hollow steel rod 100-110 cm long with a screw auger fitted on its outside to feed the chips. The cutting part of core drills has as many as 6 or 8 knives. The manufacture of core dills is quite complicated. The mass of the lightest ones, which can take a core

7-8 cm in diameter, is 20 kg or more. In some cases the sampler's body is provided with a cut-out for more convenient withdrawal of the ice core.

The drill was designed at the US Laboratory of Scientific and Engineering Investigation of Arctic Regions and is designed for surface and deep drilling holes 11 cm in diameter, and sampling firn and ice cores for analysis. The inside diameter of the sampler is 7.62 cm, the length 0.5 m. The drilling depth is down to 31 m. All the parts of the drill, for the exception of the 1 m extension pipes, are manufactured of stainless steel. The crown bit and the cutters are detachable. A screw-shaped auger is welded to the outside diameter of the drill's cylinder. Orifices are provided in the body of the drill's cylinder for the ejection of air and chips. The ice core is withdrawn through the upper parts of the cylinder, because the rod holder (a device for clamping the rods in the upper part of the drilling sleeve) is furnished with a special lock. The extension rods are aluminum pipes with adaptors and stainless steel stoppers. The drill is rotated by a windlass [170].

The drill has been comprehensively tested on wet ice under field and laboratory conditions. It was used to take ice samples from a depth of 24 m on the shelf glacier at Ellesmere Island, and on the 'ice' island T-3 from a depth of 32 m. The drill had to be pulled out to the surface in these cases by means of a special folding derrick crane and hand hoists, or by a winch with a petrol engine. The diameter of the core was 7.62 cm, its length up to 85 cm. A cut-out facilitates core sampling. A screw-shaped auger is welded from the crown bit to the cut-out. The crown bit comprises two basic working cutters with hard-alloy plates on the ends, and two additional cutters fitted in the crown but for undercutting the core and lifting it up to the surface. Owing to the extending basic cutters, the diameter of the orifice is greater than the diameter of the drill's cylinder with a welded-on auger. The additional undercutting cutters are housed in the drill's crown bit during operation in order not to hinder translational motion of the drill. At reverse rotation the band springs force the cutters to take a position perpendicular to the walls, where they make an undercut and then hold the ice core.

Core drill-sampler of Japanese ice researchers. The drill is a steel pipe with a welded auger and extending cutters in the face part. The length of the drill's body is 40 cm, and its diameter 10 cm. The drill is rotated by a brace and the drilling speed is 1 cm/s [160].

Japanese hand drill (Figure 1.55). The design of this drill is quite original. It comprises a handle-brace, a rod and a cutting part welded to the rod. The blades are fixed to semicircular discs 1 cm thick and are directed to different sides; they chip off the ice and feed the ice chips to the upper surface of the semicircular discs. The cutting part of the drill thus cuts into the ice, leaving ice chips behind. The drill is 10 cm in diameter. When it is pulled out to the surface, it acts like a piston and ejects the chipped ice. Descriptions are available of still more improved ice drills of a Swedish type, operating by planetary rotation. Experience demonstrates that drilling a hole is the least economical method; it is more economical to bore a

Fig. 1.55. Cutting part of Japanese drill [160].

hole of smaller diameter. In the case of planetary drilling the bit bores a hole.

Holes with a diameter of 100-130 mm can be drilled by means of ice drills with two knives. However, holes of greater diameter are frequently necessary, especially in thick ice.

Ice drill of the Vologda district voluntary society of hunters and anglers (Figure 1.56). This is used for drilling 170 mm diameter holes. The ice drill may be disassembled. It comprises two basic parts: a brace (type SLK-2) and a drill [33].

The drill comprises a coupling rod (seamless pipe 18 mm in diameter with walls 2 mm thick) with a bushing on the end. Two clamps of chute section are fixed to the bushing by electric welding. The clamps have stiffening ribs stamped from a 2 mm steel strip 38-40 mm wide. The clamps are welded to the bushings using a special template to prevent skewness.

A 'plate' is welded to the clamps. This is a conical disc 160 mm in diameter and 3 mm thick. It is manufactured of steel, grade ST-3. Four detachable steel knives about 50×30 mm and 2-3 mm thick, depending on the quality of the steel are fastened to the plate. At first the knives were made with three or four teeth, but now they have five or six cutting triangular teeth for better comminution of the ice. The knives are manufactured of steel, grade U-8.

The brace is coupled to the drill by means of a spline key and a lockpin with a split ring.

The ice drill has the following standard dimensions: length in assembled form 1450 mm, width (of brace) 350 mm, height (diameter of plate with knives extending further) 170 mm. The maximal drilling depth is 1000 mm, and the total mass of the ice drill is 2.5 kg.

It takes 5-10 minutes on average to drill a hole 1 m deep. It is necessary to adjust the inclination of the knives frequently by fitting a steel plate 50 mm long and 0.5-1 mm wide and thick under their rear part. The normal inclination angle of the knives to the plane of the plate is about 10°. Steep incline of the knives impedes hole drilling; a gentle slope demands more time, though it facilitates drilling.

MECHANICAL DESTRUCTION OF ICE   67

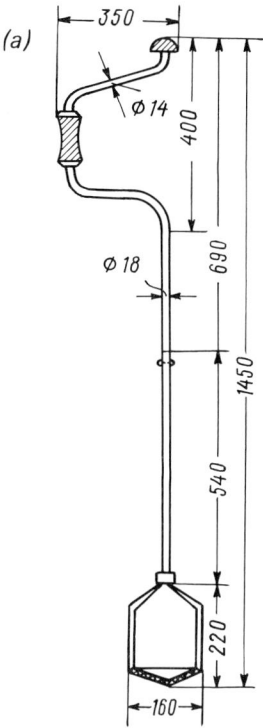
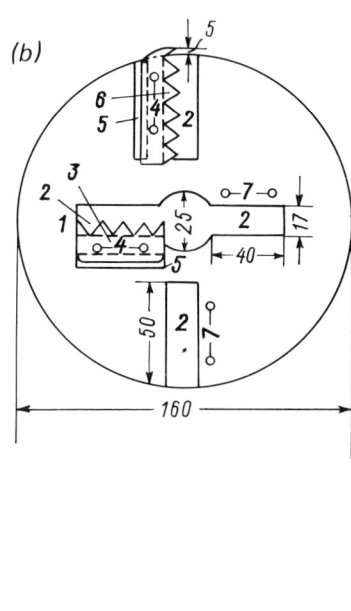

Fig. 1.56. Hand drill with four knives [33].

(a) - general view; (b) - bottom view of 'plate' (two of the four knives are shown): 1 - 'plate'; 2 - cut-out in 'plate'; 3 - detachable steel knife with five teeth; 4 - bolts; 5 - steel plate; 6 - detachable steel knife with six teeth; 7 - holes for bolts.

The drill rotates counterclockwise.

Trepanning ice drill PI-8 (Figure 1.57). This drill has been developed at the Arctic and Antarctic Research Institute by N.V. Cherepanov, and is used widely in practice (Author's Certificate 472,237 USSR). It is a new type of core drill with ring diameters of 180, 220 and 320 mm [116, 131, 178]. The drill is designed for hand-drilling holes of various diameters in ice with core sampling. It is unlike existing types of core drills in the absence of a metal sleeve and an auger, which are the basic components of core drills.

The trepanning drill comprises a massive metal ring (6), whose bottom part is a sharp face knife of wedge-shaped section, ensuring stable position at rotation and automatic adjustment of the cutting speed.

A groove (5) at an angle of 40-45° and 25-30 mm wide is cut in the ring to secure the knife. The groove ensures free removal of the chips to the upper flat surface of the ring. A toothed cutter is fastened on

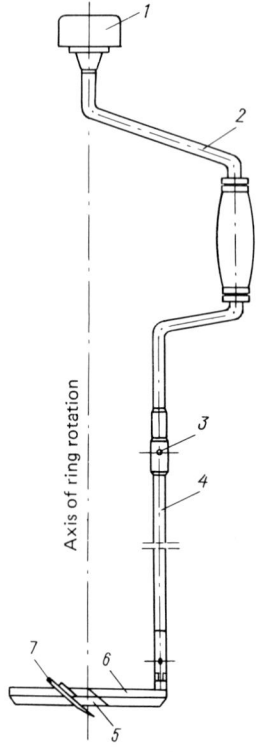

Fig. 1.57. Trepanning drill [178].

1 - brace handle; 2 - brace; 3 - catch with lockpin; 4 - rod; 5 - groove; 6 - ring; 7 - toothed cutter. The axis of ring rotation is shown.

one of the groove's sides, and its position is adjusted by two locking screws. The ring is coupled to the brace (2) by means of the rod (4), catch and lockpin (3).

The rod and brace are manufactured of hollow pipes to lighten the drill. An increase in the diameter of the ring without an essential increase in the drill's total mass makes it possible to drill a hole of any size in ice.

The total length of the drill is 140 cm, but its design makes it possible to use additional rods to drill holes 3-4 m deep.

The following conditions should be observed to ensure reliable operation of the trepanning drill:
1. the rod (4) should be arranged strictly perpendicular to the upper plane of the ring (6);
2. the brace's upper guiding handle (1) should coincide with the axis of ring rotation and its center;
3. the cutter (7) should project 2-3 mm under the bottom guide face knife;

4. the cutter's point in working position should be 5-6 mm wider than the ring, and the bottom part of the cutter should project 2-3 mm on each side of the ring. The upper part of the cutter should not project beyond the ring's edge and scratch the hole wall or the core.

If the first precaution is ignored the rotation of the rod is gradually impeded owing to friction against the hole wall or the core when drilling at a depth of 40-60 cm or more. If the angle of the ring plane to the rod is more than 90°, the rod presses to the wall; if the angle is less it presses to the core.

These latter faults can be eliminated as follows under field conditions: a new 3-4 cm deep hole is drilled, a ring is inserted and slightly frozen-in, then the rod is straightened by slight pressure, its position being controlled by a plumb line and level, or a triangle.

Misalignment of the brace handle centre with the axis of ring rotation is associated with slipping of the ring at rotation and shearing of the core's upper layers. When the ring sinks further into the ice, the rod of the drill and the handle of the brace vibrate strongly, impeding the drilling process. Insignificant (up to 3-5 cm) deviations of the brace handle from the ring's centre may be eliminated by bending the upper or bottom shoulder of the brace. The brace should be replaced if the deviations are too great.

Sometimes insufficient fastening of the lock screws may disturb the correct position of the cutter (7), and it may be displaced up or down. At a small clearance the drill rotates easily without sinking; vice versa, an increase in the clearance greatly impedes rotation of the brace. Three transverse slots in the ring's groove are designed for adjusting the position of the cutter and fastening the latter, and there is a corresponding projection on the cutter which is fastened by two lock screws.

Faults in the operation of trepanning drills are associated with failure to observe the fourth condition mentioned above. The drill operates easily only at the beginning, at the surface. When the ring sinks 1-2 cm, it rotates with great difficulty and the drill no longer sinks. Before replacing the cutter, it is necessary to measure the clearance between the core and the ring, as well as the clearance between the ring and the hole wall. The clearance should be at least 2-3 mm on either side of the ring. A smaller clearance impedes the operation, because the ring either seizes, or runs idle if it sinks on the core. In this case it is necessary to replace the cutter.

The quality of trepanning drill operation depends on the physico-mechanical properties of the ice: the drill sinks more easily into sea ice than into freshwater ice under the same temperature conditions.

The cutter should be very sharp, and the clearance accurate even with drop of the temperature, when the ice hardness increases greatly.

Unlike other drills, the trepanning drill does not require comprehensive practical experience and great physical effort. The work can be easily done by one man even when drilling holes up to 320 mm in diameter.

In order to put a trepanning drill into working condition it is necessary to connect the brace's handle to the rod, the head of the former being in one line with the ring's axis of rotation. The brace is fastened by a catch and a lockpin, and the cutter is set in the working position. To do this, it is necessary to slacken the lock screws, move the cutter 2-3 mm over the bottom plane of the ring and tighten the cutter with the screws.

Before starting operation, clean an even ice site of snow, fit the ring on the ice surface and rotate the brace slowly counterclockwise, keeping the rod vertical. The brace's upper handle (1) should always be in the centre of the ring's axis of rotation. Increase the pressure on the brace's handle while the ring sinks into the ice. When the ring sinks 5-10 cm, withdraw the drill and remove the chips. The latter operation is repeated at every 5-6 cm depth. Accumulation of a large quantity of chips should be prevented because it impedes free rotation of the rod. Compacting chips may cause ring seizure in the hole.

The core should not be broken off until the drilling operation is completed. It impedes the removal of chips, especially when the chips are dry and are not kept on the ring. If the core breaks accidentally, it should be withdrawn immediately. The removal of chips is improved by introducing a solution of common salt into the hole in order to wet the chips and make them viscous. The chips are then easily held on the ring. The drill itself should not be used to break and withdraw the core from the hole, because this damages the rods of the drill; which are manufactured of thin-walled pipes. If it is necessary to break off the core, this is done by a light impact on a wooden wedge, which is inserted into the circular groove, and the core is withdrawn by a special rod fastened on to a spare one. If the ice thickness is more than 1 m, the main rod and a spare rod are used, first removing the brace, fitting a rod in its place, and turning the drill upside down with the ring upwards.

Drills designed at N.E. Bauman Moscow Higher Technical School (Figure 1.58). Experimental drills, designed at the Moscow Higher

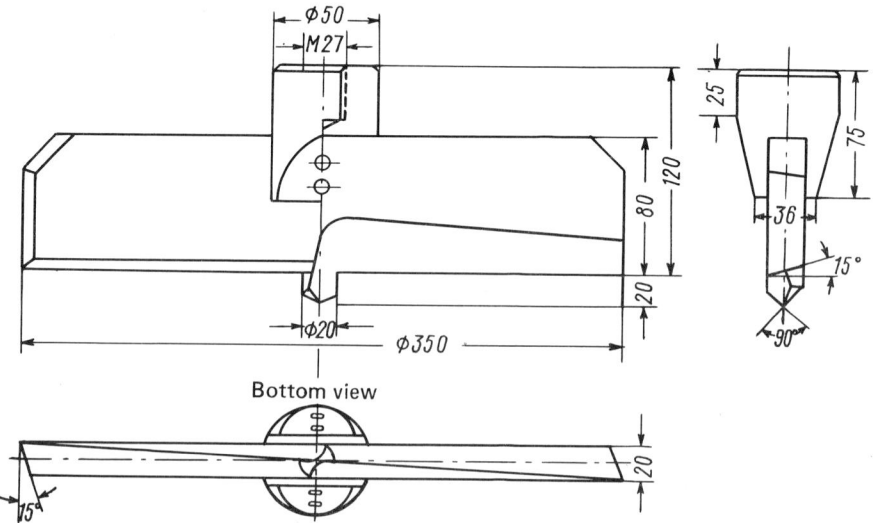

Fig. 1.58. Drill designed at N.E. Bauman Moscow Higher Technical School.

MECHANICAL DESTRUCTION OF ICE

Technical School, with a diameter of 250 and 350 mm, consist of a flat bit with a horizontal cutting edge directed to the centre of the drill. Beside the main cutting edge, the drill has lateral and upper cutting edges to assist withdrawal of the drill from the hole in case it is displaced in relation to the hole.

The drill has a centre bit. The bit's diameter is 26 mm in a 250 mm drill, and 20 mm in a 350 mm drill. The centre's jumper is 2 mm in a 250 mm drill, and 4 mm in a 350 mm drill.

GEOMETRY OF DRILL

Side rake angle (α_1)	15°
Face rake angle (α_2)	10°
Lip angle (β)	75°
Rake angle (γ)	0°
Cutting angle (δ)	90°

Three-blade drill designed at All-Union Research Institute of Marine Fishing and Oceanography (Figure 1.59). The drill is a three-blade unit with interchangeable knives and a burster. The cutting knives

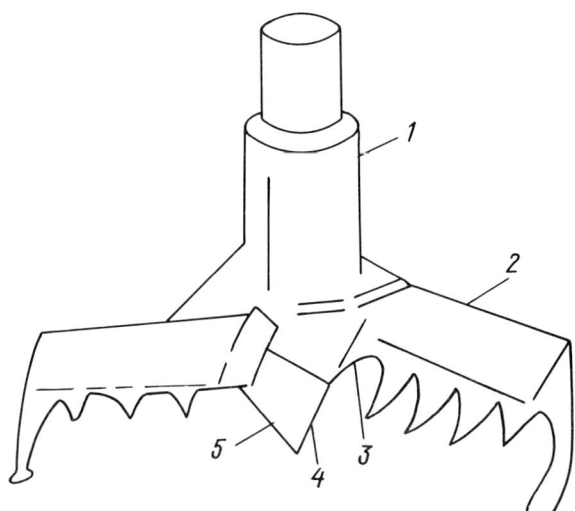

Fig. 1.59. Principal view of three-blade drill designed at All-Union Research Institute of Marine Fishing and Oceanography [147].

1 - adapter for fastening ice drill to spindle; 2 - cutting knives; 3 - fastening nuts; 4 - shaped holder; 5 - burster.

may be arranged at various cutting angles. The drill's diameter may be 250, 310 or 390 mm, depending on the knife length. The drill comprises a shaped holder, on which three knives are fastened by means of a cone and two nuts. A three-blade burster is fastened on a thread below the holder. The bit is screwed on by means of an adapter.

SPECIFICATIONS OF THREE-BLADE DRILL

Drill diameter (mm)	250,	310,	390
Number of teeth			
cutting	13,	16,	16
undercutting	3,	3,	3
Length of tooth cutting edge (mm)			
main	5,	5,	5
undercutting	8,	8,	8
Length of cutting edge (mm)			
total, without burster	74,	89,	89
total	224,	239,	239

The drill's teeth are arranged in a helix in staggered order from the centre to the periphery. The tooth nearest to the centre is the most sunk (longest) one. The next tooth is on another blade. It is displaced to the periphery in relation to the previous tooth, and is less sunk. The next (third) tooth on the third blade is also displaced to the periphery and is sunk even less, and so on along a helix successively on each blade up to the extreme peripheral tooth.

The gradual reduction in the sinking of the teeth creates an angle of $2\varphi = 170°$ at the point of the bit on rotation of the drill.

The use of a percussion rotary cutter (of the conical-bit type) in a hand tool has been suggested, ensuring contact with the material at the required surface by preliminary shearing of the ice on cutter rotation, causing the cyclic development of separation cracks (Figure 1.60). This is advantageous from the power consumption point of view [7].

The bit, passing through the centre of the tool, is an auger to remove chips from under the tap part, cutting a slot for a screw by means of vibratory impacts with an increasing lead. A screw with an increasing lead separates large pieces of ice. They are removed by pneumatic transport through a mill which comminutes the ice. An unbalanced vibration exciter energizes the tap and the screw [7].

Ice drill of Tomsk Fishing Trust (Figure 1.61). This is the most convenient, reliable and economical drill currently in use, especially if the volume of work is small.

SPECIFICATIONS OF TOMSK ICE DRILL

Maximal drilling depth (mm)	900
Hole diameter (m)	320
Drilling speed (m/min)	0.6
Time of hole drilling at ice thickness 70-80 cm (including time to move to another hole) (min)	2-3
Drill feed mechanism	manual
Engine power (at 5200 r.p.m.) (kW)	2.6
Ice drill mass (without sledge) (kg)	60

MECHANICAL DESTRUCTION OF ICE

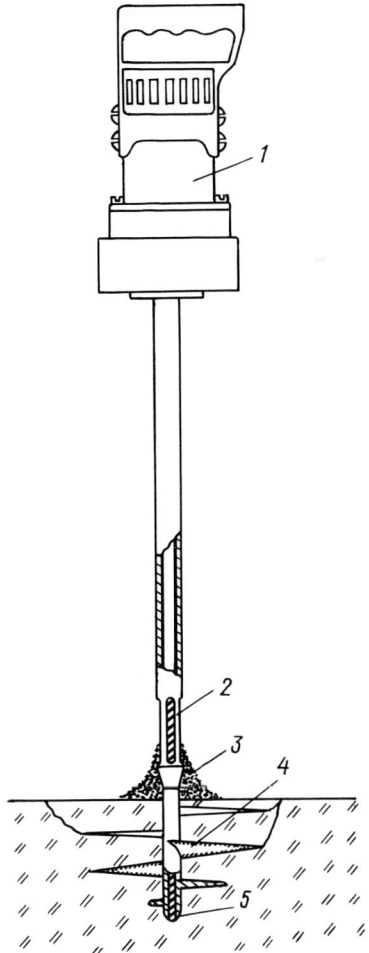

Fig. 1.60. Mechanical hand-vibration impact drill [147].

1 - unbalanced virbation exciter;
2 - auger; 3 - tap; 4 - screw with increasing lead; 5 - bit.

The drill is operated by two workers: drilling is safe at a minimal ice thickness of 20 cm.

The ice drill comprises Druzhba-60 petrol saw engine, an additional reduction gear with control levers, bit guides, bit spindle (rod), bit proper, metal bearing plate, and sledge. The ice drill is mounted on the rear of the sledge on the metal bearing plate.

The bit comprises a hexahedral hub with a burster screwed on its end and two welded holders for fastening the rake and knife. The latter two parts are detachable.

When the bit rotates, the rake loosens the ice thickness, and the knife with a sharp rectilinear cutting edge shears off the loosened ice.

The ice drill on the sledge can be shifted from working to transport position (the transport position is shown in Figure 1.61 by dotted lines).

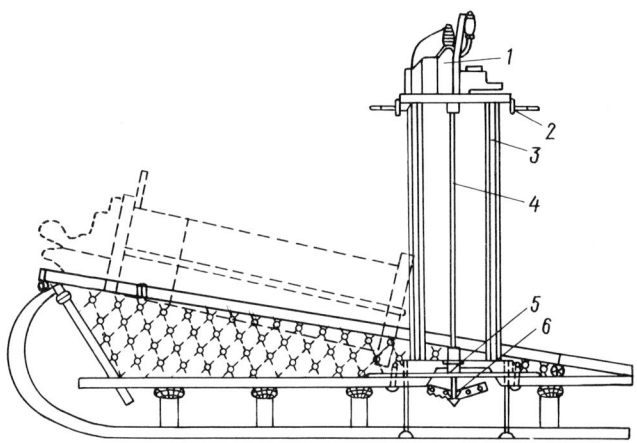

Fig. 1.61. Ice drill designed at Tomsk Fishing Trust [145].

1 - engine of Druzhba-60 petrol saw; 2 - control levers; 3 - bit guides;
4 - 'spindle'; 5 - bearing plate; 6 - bit.

To make holes, the ice is first drilled to a depth of 25-30 cm, then the drill is pulled out, partially cleaning the hole of ice chips; the drill is sunk again and the next 15-20 cm of ice drilled; the drill is pulled out again, and so on. The thicker the ice, the more difficult it is to drill its bottom layers, and it becomes necessary to pull out the drill more frequently. To facilitate the operation after drilling the ice to a depth of 45-50 cm, it is advisable to pull out the drill several times to remove all the chips from the hole.

It is prohibited to drill ice more than 30 cm deep in one pass, because this causes bit seizure in the hole and may cause breakdown of the machine.

The ice drill of the Tomsk Fishing Trust replaces about 10 manual workers.

The cost of drilling a hole in ice 80 cm thick is 40 kopecks, i.e. about one-third of the cost invovled when using common ice picks [145].

Mechanical drill designed by US Army Research Laboratory of Arctic Regions (Figure 1.62). This is a unit 6 m long with a mass of 100 kg. The upper third of the drilling complex houses a motor and a transmission mechanism, the central part comprises a container for ice chips, and the central shaft and cutting crown bit are in the bottom. Three-phase current (380 V) is supplied via a braided steel cable from a 1.5 kW electric motor to the 2 m long central shaft, rotating inside a stationary external pipe. The cable's length is 7.2 m, its diameter 22 cm and its mass 0.75 kg/m. The winch is driven by a 5 kW electric motor via a belt transmission. The average velocity of up-and-down motion of the drill is 25 m/min. Two cutters on the cutting crown bit in the bottom part of the shaft cut a circular furrow 1.5 cm wide; the ice chips move up along the helix of the auger and accumulate in the

MECHANICAL DESTRUCTION OF ICE

container, which is emptied after each drilling cycle.

The core length drilled per cycle is 2 m at a diameter of 90 cm. The total hole depth drilled in glacier ice in Iceland was 415 m [194].

Ice-drilling unit designed by Fershtut and Pashchenko (Figure 1.63). The unit is a mobile non-self-propelled plant, comprising a 4.4 kW petrol engine (model L-6) for the drilling and hauling assemblies. All the units are mounted on metal sledges of welded design.

SPECIFICATIONS OF FERSHTUT AND PASHCHENKO ICE-DRILLING UNIT

Purpose	To drill holes in ice and haul a seine
Ice drilling depth (mm)	<400
Hole diameter (mm)	300-350
Drilling speed (depending on specific conditions (m/s)	5-10
Hauling force (kg)	670
Hauling speed (at engine rated r.p.m. (m/min)	26.5
Hauling speed adjustment	control of fuel supply to engine
Engine	petrol (model L-6)
Engine power (kW)	4.4
Rotational speed (r.p.m.)	
engine	2200
drilling spindle	1100
warping drums	55
Number of warping drums	2
Warping drum diameter (mm)	170
Mass of unit (kg)	350

The sledge has a platform for the driller and a device for fixing the unit on ice. The unit drills ice 50-60 cm thick, but it operates normally only at an ice thickness up to 30 cm. When drilling deeper, the spindle vibrates violently and great effort is required to feed the bit.

When the engine is started, the unit operates idle

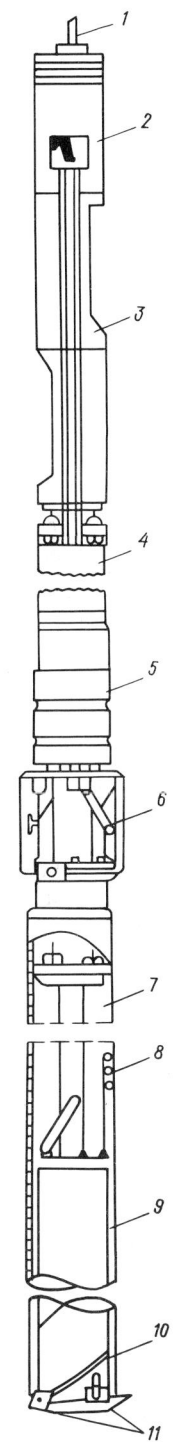

Fig. 1.62. Sectional view of American mechanical drilling unit [194].

1 - cable; 2 - suspension; 3 - orientation block;
4 - electric motor; 5 - transmission mechanism block;
6 - union; 7 - ice chip container; 8 - gate;
9 - central shaft; 10 - internal joint safety device;
11 - bit cutters.

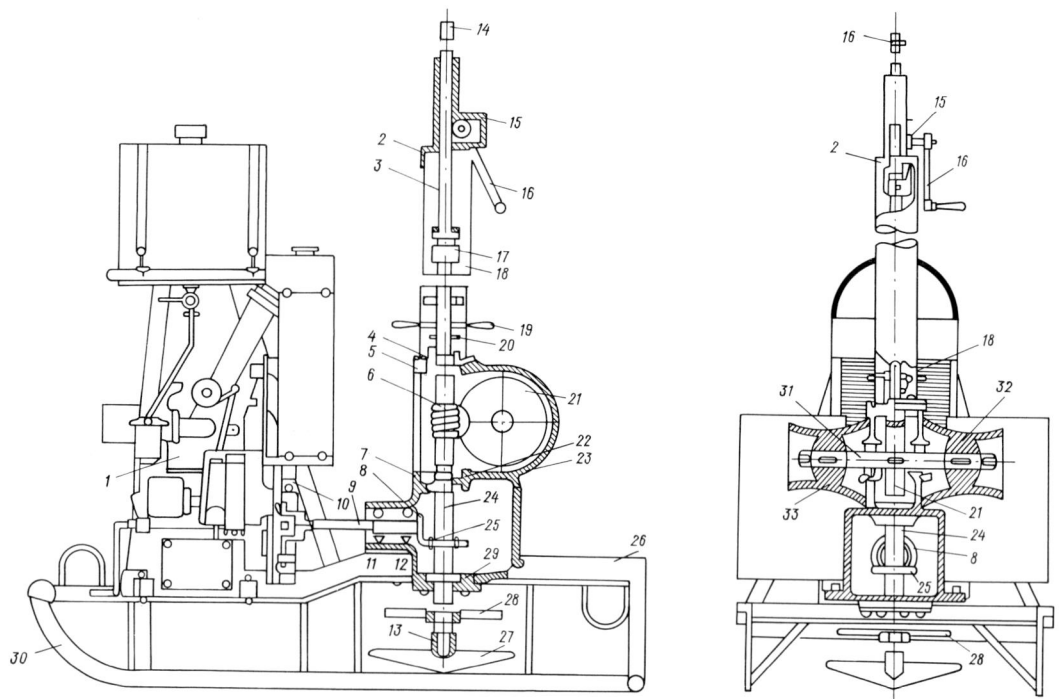

Fig. 1.63. Ice drilling unit designed by Fershtut and Pashchenko [147].

1 - engine; 2, 17 - sleeve; 3, 14 - pressure gauge journal; 4, 20 - half jaw clutch; 5, 11, 12, 22, 23, 29 - bearings; 6 - shaft; 7 - reduction gear housing; 8 - driving bevel gear; 9 - roller; 10 - coupling; 13 - spindle key slot; 15, 25 - gear; 16 - feeding mechanism handle; 18 - pipe; 19 - handle; 21 - worm gear; 24 - driving hub; 26, 30 - sledge; 27 - bit; 28 - snow remover; 31 - shaft; 32, 33 - single-stage warping drums.

for 10-15 min before hole drilling is commenced. The unit is operated by a motor-man and a driller. The unit is brought to the drilling site, the driller gets up on the fixing platform and starts the drilling cycle by turning the handle of the feeding mechanism. At the end of the drilling cycle (when the feed force drops sharply) the driller lifts the drill to the initial position, fixes it by means of a lock to prevent spontaneous sinking, and gets off the fixing platform. The unit is automatically freed from the fastening, and can then be moved to the next hole drilling site.

The time required to prepare for drilling is 20-25 s, including moving the unit from one hole to another. Thus, the time required for drilling one hole is about 1 min at the given ice thickness. The time required for drilling one hole manually is 2-3 min.

Fershtut-Pashchenko ice-drilling units have certain serious disadvantages: the bit feeding mechanism is imperfect, and the rotational speed of the bit is as high as 1100 r.p.m. When the bit reaches the water, there is an instantaneous outburst to the surface that impedes operation. Also, the diameter of the warping drum (170 mm) is insufficient.

Sixteen different drills have been tested, including drills designed by G.N. Izmailov and I.A. Kormin (Figure 1.64a), G.M. Brovkin (Figure 1.64b), drills designed at the Barabinsk Fish Plant of the Novosibirsk City Fishing Trust, etc.

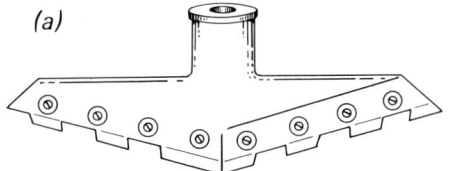 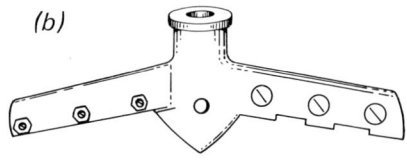

Fig. 1.64. Drills designed by (a) G.N. Izmailov and I.A. Kormin, (b) G.M. Brovkin [147].

The best test results have been obtained from drills of G.M. Brovkin and the Barabinsk Fish Plant. The latter is based on the principle of breaking out pieces of ice. The chips formed on drilling are large, the bit is fed without great force, and the chips are easily ejected when drilling ice up to 300 mm thick. It is possible to drill 500 mm deep holes.

LB-1 ice-drilling unit (Figure 1.65). A new type of an ice-drilling unit has been designed at the State Institute for Designing Enterprises of Fishing Industry, on an assignment from the All-Union Research Institute of Marine Fishing and Oceanography, using the operational experience of the first series of ice drilling units designed by Fershtut and Pashchenko.

SPECIFICATIONS OF LB-1 ICE-DRILLING UNIT

Purpose	To drill holes in ice and haul a seine
Maximal drilling depth (mm)	600
Hole diameter (mm)	<350
Bit rotational speed (r.p.m.)	665
Drill feed	manual
Working tool of hauling device	two warping drums
Warping drum diameter (mm)	225
Hauling speed (m/min)	12 and 23
Hauling force (kg)	<1000
Drive	L-6 engine
Engine power (kW)	4.4
Rotational speed (r.p.m.)	2200
Mass of unit (kg)	470

Overall dimensions (mm)
　length   1330
　with collapsible platform   1710
　width   932
　height   1526

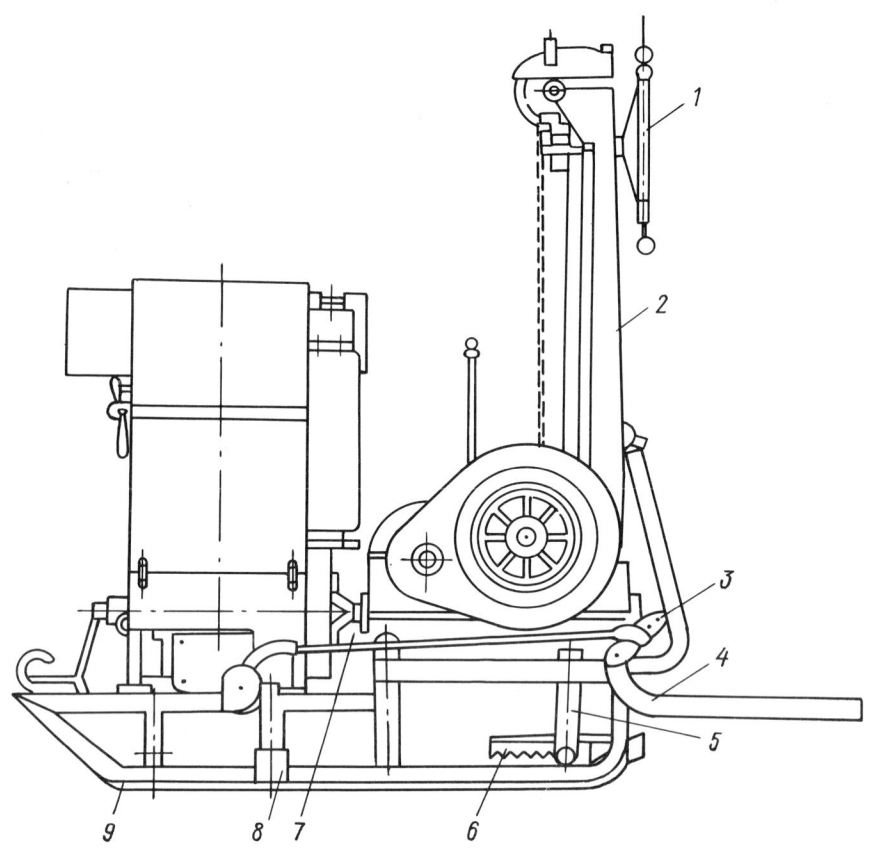

Fig. 1.65. Ice-drilling unit LB-1 [147]

1 - steering wheel; 2 - post; 3 - levers; 4 - collapsible platform; 5 - bit; 6 - bit blade; 7 - tie-rods; 8 - rests; 9 - sledge.

The ice-drilling unit is a non-self-propelled mobile plant mounted on a metal welded sledge.

The unit has certain advantages over Fershtut and Pashchenko's earlier version. The bit and warping drums disconnect more reliably by means of friction couplings. The bit feed mechanism is also more reliable; the height of the unit and the spindle length are less. A collapsible platform facilitates access to the bit.

The unit comprises a drilling tool (Figure 1.66). The bit blades

MECHANICAL DESTRUCTION OF ICE

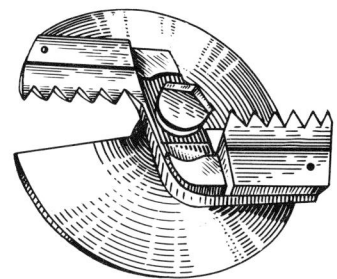

Fig. 1.66. Bit of ice-drilling unit LB-1 [147].

contribute to better ejection of the chips from the hole, and holes are self-cleaning on drilling to a depth of 250-300 mm. When drilling to a greater depth, it is necessary to clean out the hole once or twice by pulling out the bit.

The chips are withdrawn to the ice surface by the bit blades and by the effect of the spreader's centrifugal force.

Ice drill designed by N.E. Shlyaev (Figure 1.67). The ice drill

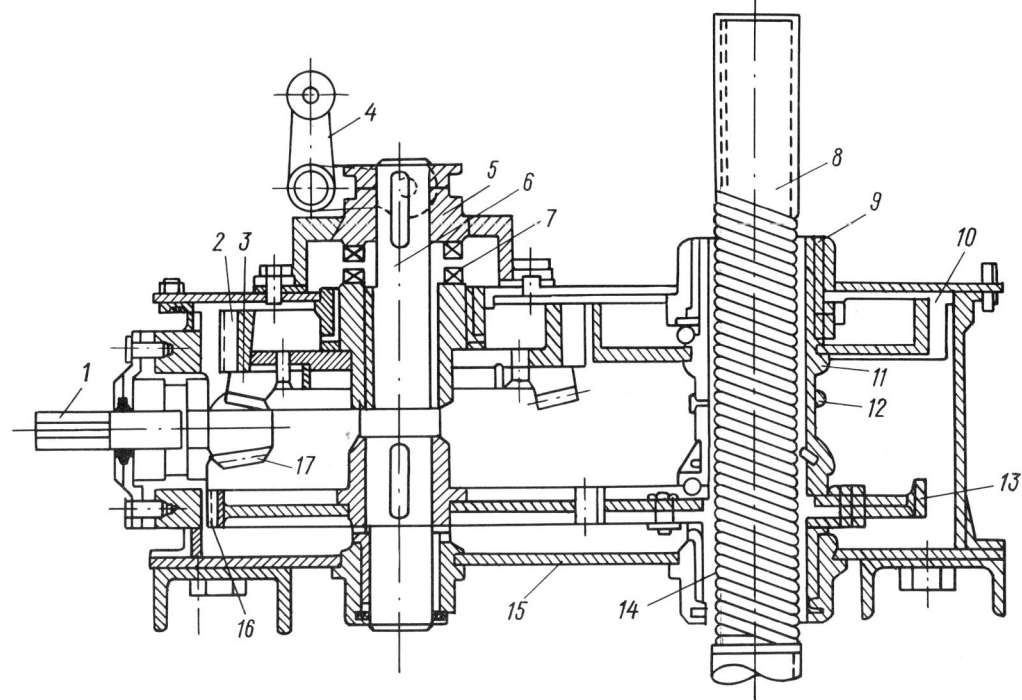

Fig. 1.67. Ice drill designed by N.E. Shlyaev [147].

1 - shaft; 2, 16 - cylindrical gears; 3 - bevel gear; 4 - lever; 5 - half jaw clutch; 6 - shaft; 7 - jaw clutch; 8 - lead screw (ice drill spindle); 9 - sleeve; 10, 13 - gears; 11 - key; 12 - screws; 14 - nut; 15 - reduction gear bottom cover; 17 - toothed bevel gear.

has been designed for Siberia, the Northern USSR and the Far East, and is mounted on the KD-35 tractor.

SPECIFICATIONS OF SHLYAER ICE DRILL

Ice drilling depth, including ice cover (mm)	1400-1500
Hole diameter (mm)	<400
Bit feed	mechanical
Drilling speed (mm/s)	70
Bit rotational speed (r.p.m.)	350
Drive	KD-35 tractor
Overall dimensions of gearbox (mm)	
length	800
width	400
height	400

The drilling mechanism is mounted in the rear part of the tractor on a special platform which has the configuration of a sledge. The bit is fixed on a spindle and comprises four blades with interchangeable cutting plates fixed on two of them. The plates are provided with cuts of variable pitch. The blades are a part of the auger. The diameter of the bit is 400 mm. A burster 160 mm in diameter is fastened to the bottom end of the bit. The burster bores into the ice first, ensuring a stable position of the main bit.

Ice chips are not ejected from the hole during the drilling process. The chips are removed only when the bit is pulled out, and simultaneously a little water is ejected from the hole. The ice chips are granulated by the drilling operation.

The ice drill is operated from the tractor cab by the driver, who is also the driller.

An ice drill of this design has been tested on Lake Chany. The ice sheet was 700-1050 mm thick, the snow cover was 250-400 mm, and the air temperature was from -10 to -18°C. About 400 holes were drilled during the tests.

Stop-watch studies demonstrated that the time of drilling a hole (including back stroke of the bit) varied from 28 s to 35 s, depending on the ice thickness within the limits stated above, and the time for drilling a hole was 56-70 s, including the time required for the tractor to move from one hole to another (to a distance of 20 m), engagement and disengagement of the drill.

The ice drill can be used to cut lanes. A hole is drilled more than 10 times faster than by manual labour.

The ice drill of N.E. Shlyaev is used to drill holes 1400-1500 mm deep. However, the use of the drill is limited because it is mounted on a KD-35 tractor, which demands a lot of fuel. It is also necessary to have a tractor driver to operate it.

The total mass of the tractor with the ice drill makes it possible to use it only if the ice is thick (30 cm). The drill is disassembled from the tractor when the ice begins to melt, which limits the time during which it can be used.

MECHANICAL DESTRUCTION OF ICE

Non-self-propelled ice drilling unit, model 26 (Figure 1.68). R.I. Pshenichnikov and I.N. Morozov have developed an ice drilling unit, model 26.

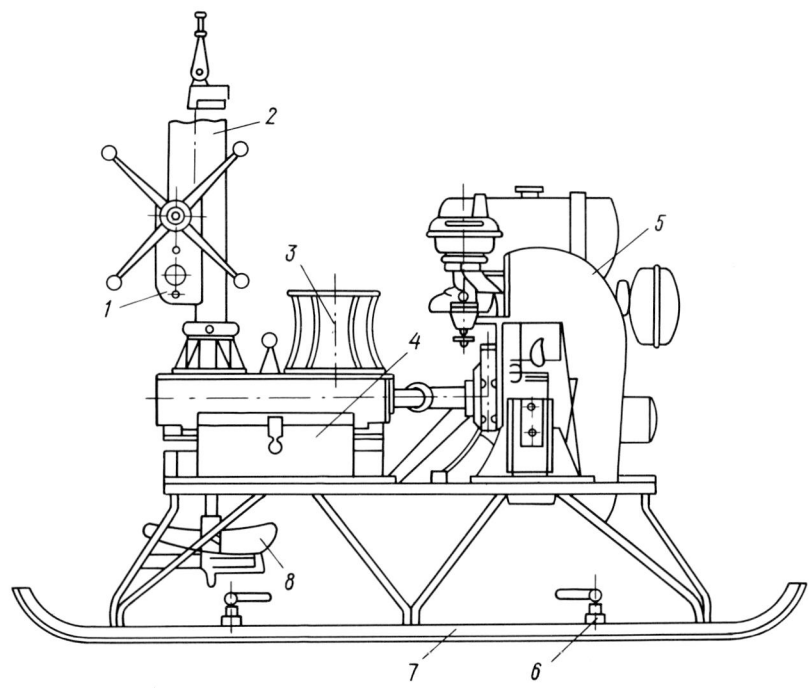

Fig. 1.68. Non-self-propelled ice-drilling unit, model 26, designed by R.I. Pshenichnikov and I.N. Morozov [53].

1 - feed mechanism; 2 - drill column; 3 - warping drum; 4 - reduction gear; 5 - engine; 6 - catch; 7 - sledge; 8 - bit.

The motor ice drill is a non-self-propelled mobile unit on a metal sledge. It is designed to drill (bore) holes down to 1250 mm deep with a diameter of 320 mm.

The motor ice drill comprises an engine (model ZID-4.5, series A), a frame, reduction gear, feed rod with steering wheel, cutting tool (bit), and warping drum (capstan).

The drill (Figure 1.69) comprises a steel hexahedral framework with a welded auger blade to eject ice chips, to which are fastened three cutting knives. The bit can be sharpened rapidly because the knives are detachable. The knives are secured on three blades arranged circumferentially at 120°. Each blade with its knives is arranged at 55° to the veritcal axis of the framework. A replacement knife is bolted to the blade of the bit. The knives of the bit have different cutting

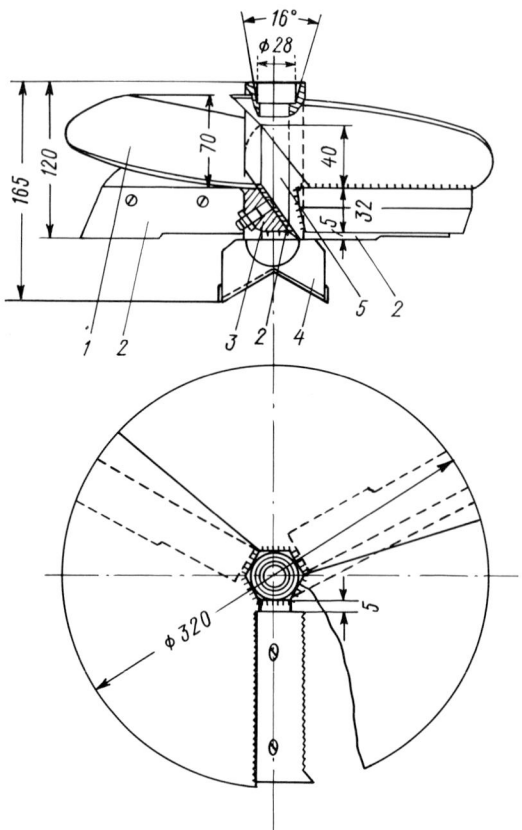

Fig. 1.69. Ice drill model 26 [53].

1 - auger blade; 2 - knives; 3 - blade; 4 - flat drill; 5 - framework.

edges. The edge of one knife is shaped like a toothed rake of trihedral section, enabling it to shear the ice. The second knife has a protruding external cutting edge to chip off the ice further from the centre. The third knife chips ice under the central part of the bit by its protruding internal cutting edge. The knives are manufactured of tool steel, grade U-8.

A two-bladed flat drill is tied to the bottom part of the framework. The tips of the flat drill's knives extend below the centre. The bit diameter is 314 mm.

When drilling holes in ice, the motor ice drill is operated by the reduction gear control arm and the steering wheel of the bit feed mechanism.

The imperfection of the bit design, the great force of feed required, and displacement in drilling owing to unsatisfactory design of the screw catches, are the disadvantages of the unit.

MECHANICAL DESTRUCTION OF ICE 83

The unit is designed mainly to mechanize the process of fishing under ice in the central regions of the USSR. It has been operated in Siberia with N.E. Shlyaev's bit to accelerate the drilling of holes. The drilling speed reaches 0.5 m/min, depending on the specific conditions.

OLB-42 light ice drill (Figure 1.70). This has been in production since 1965. Unlike model 26, it is designed to drill ice in regions

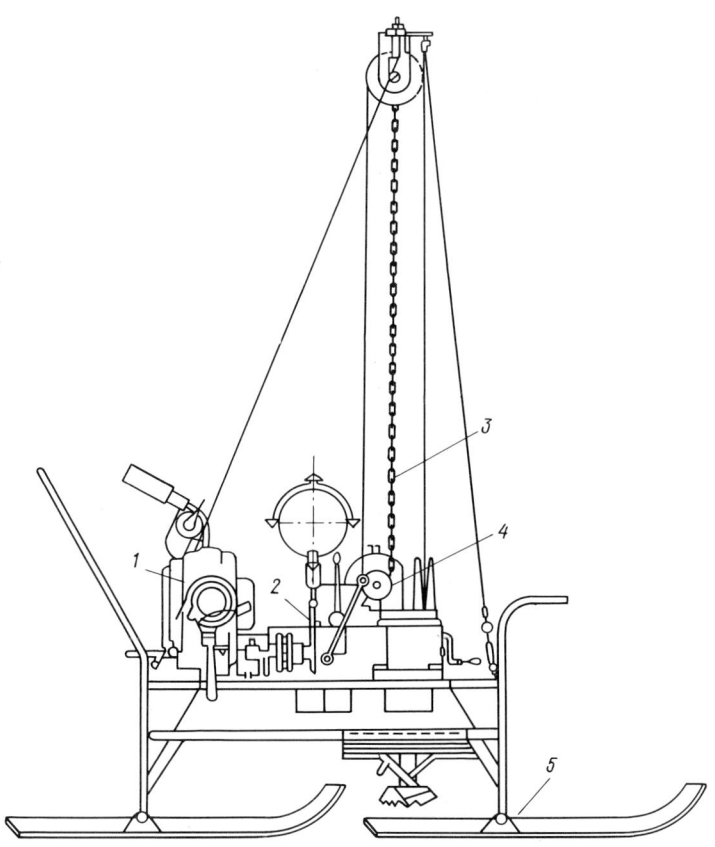

Fig. 1.70. Light ice drill OLB-42 [147].

1 - engine; 2 - reduction gear; 3 - drill column; 4 - feed mechanism; 5 - sledge.

where the ice thickness is 600-800 mm and the snow cover is at most 400 mm. The ice drill is designed with mechanical bit feed, and its accelerated withdrawal facilitates the labour involved. The ice drill comprises an engine, reduction gear, drill column and feed mechanism. All the assemblies are mounted on a sledge. The ice drill is a non-self-propelled unit, like model 26.

The unit comprises a 3.7 kW petrol engine (model VP-150) from the Vyatka motor-scooter. The gearbox produces three rotational speeds: 195,

325 and 520 r.p.m. The three rotational speeds of the bit and the three feed speeds are a great advantage, because it is possible to select the optimal drilling regime in each case.

The mechanism is designed so that the bit can be fed and pulled out either when it is rotating, or when it is in neutral. (Figure 1.71).

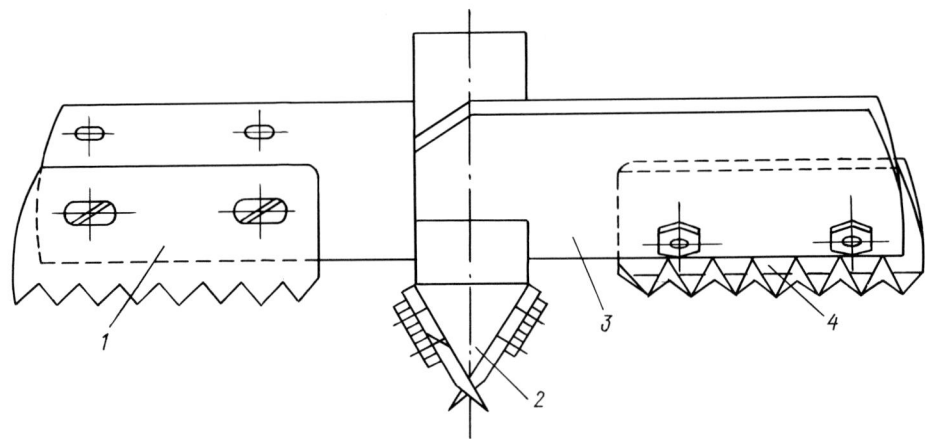

Fig. 1.71. Bit of light ice drill OLB-42 [147].

1, 4 - right-hand and left-hand rakes; 2 - burster; 3 - body.

An extending auger is used in the ice drill to remove the chips from the hole. One end of the auger is fastened to the bit, and the other to the sledge.

The chain rotating the bit and the one feeding it can be engaged independently of one another. This is an important feature because it is sometimes necessary to pull out a non-rotating bit from the hole to prevent water pouring out on to the ice surface. The mass of OLB-42 ice drill is 116 kg. It is readily mobile because it is mounted on a sledge with wide runners and the specific pressure on the snow is therefore rather low. These ice drills are widely used for fishing under ice in some regions of the USSR.

Tractor-mounted ice drilling unit (ILB-1, ILB-2). The drill of the first model is driven by the tractor power takeoff, and the bit is fed and withdrawn mechanically via a system of cylindrical and bevel gears. The bit's rotational speed and its feed are constant and depend only on the rotational speed of the tractor power takeoff. Tests of this ice drill have demonstrated that it can drill ice more than 1000 mm thick, and the time for drilling one hole is 10-15 times less than in the case of hand drilling. However, operational experience has indicated that one constant feed of the bit and one rotational speed is insufficient for normal operation because the mechanical characteristics of the ice are known to change with temperature.

N.E. Shlyaev developed a new variant of the ice drill with hydraulic cutting tool feed. Though the rotational speed of the bit remained constant at 272 r.p.m., it became possible to adjust the bit feed because it was possible to vary the speed of sinking and withdrawing the spindle with the bit.

SPECIFICATIONS OF ILB-2 ICE-DRILLING UNIT

Drive	tractor power takeoff
Maximal drilling depth (mm)	1500
Hole diameter (mm)	350
Bit rotational speed (r.p.m.)	272
Hauling member	warping drums
Hauling speed (m/min)	15
Mass of tractor, drilling part, winch (kg)	5000

The drilling part of the ILB-2 ice-drilling unit is a mechanism mounted at the rear of the tractor (Figure 1.72), driven by the tractor power takeoff. The drilling part is fastened to the tractor on special brackets. A winch, arranged in front of the tractor, is driven by the power takeoff via a transmission system. The winch comprises two horizontal warping drums mounted on the load shaft. The drilling part and the winch are controlled by the driver from the tractor cab.

An automatic device is available to adjust the drilling depth from 1000 to 1500 mm.

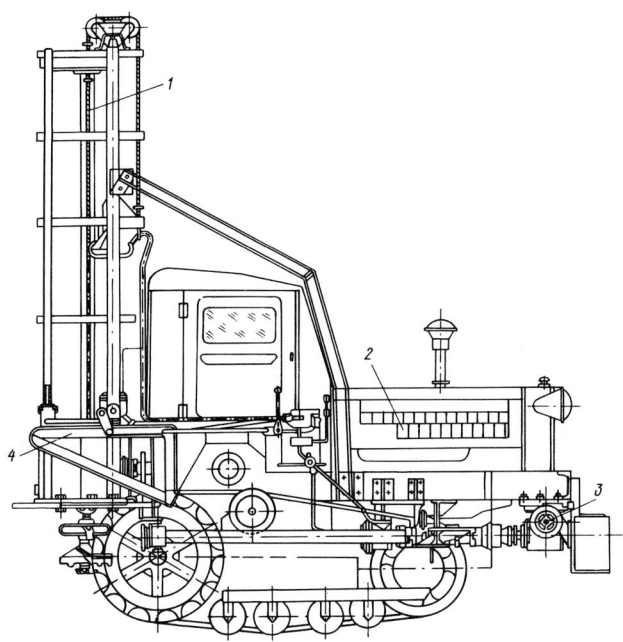

Fig. 1.72. Tractor-mounted ice-drilling unit ILB [150].

1 - drilling mechanism; 2 - tractor; 3 - winches; 4 - brackets.

When drilling holes, the unit is operated by the tractor driver. The average time for drilling a hole at an ice thickness of 900-1000 mm with snow cover up to 270 mm is 15 s; the average time required to pull out the bit is 20-25 s, and the average total time for drilling one hole, including auxiliary operations and driving the tractor 20 m from one hole to another, is 60 s. This is 10-15 times less than the time required for hand drilling.

Tractor-mounted ice-drilling units ILB-2 have certain disadvantages which must be set against their considerable advantages. The unit's centre of gravity is too high, which lessens its stability and creates serious difficulties in moving across rugged terrain. The frame of the unit and the mobile carriage lack sufficient rigidity, which sometimes results in deformation.

Self-propelled ice drilling unit PRAG-GPI-56 (Figure 1.73). This unit has been developed on the basis of the GAZ-47 caterpillar conveyer, which is a cross-country vehicle designed for transporting people and cargo in the northern regions of the USSR. The engine's power is 54.5 kW at a crankshaft rotational speed of 3000 r.p.m.

The unit comprises a conveyer, drilling device, and winch. The conveyer is 5200 mm long and 2435 mm wide and its mass is 3650 kg. The fuel consumption is 40-100/100 km, depending on the road conditions. The capacity of the fuel tank is 250 l.

The drilling device is arranged in the rear of the machine and comprises a drive from the conveyer's engine, a cutter head, column, spindle, bit, bit feed, and a withdrawal mechanism with a hydraulic

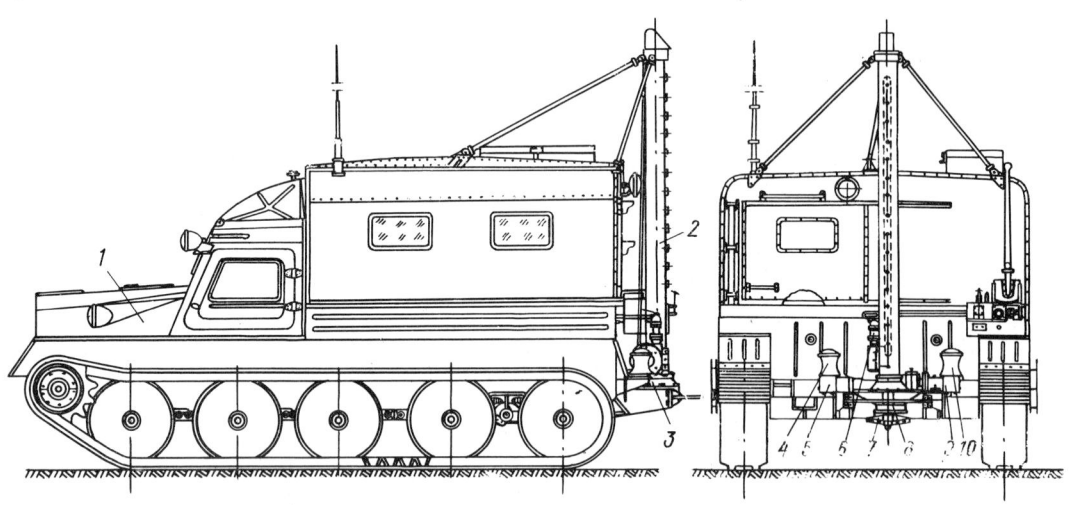

Fig. 1.73. Self-propelled ice-drilling unit PRAG-GPI-56 [147].

1 - conveyer; 2 - column; 3 - cutter head; 4, 10 - warping drums; 5, 9 - reduction gears; 6 - bit feed and withdrawal mechanism with hydraulic motor; 7 - bit; 8 - spindle.

MECHANICAL DESTRUCTION OF ICE

motor. An auger is arranged between the bit and the cutter head to remove the chips from the hole. The design of the drilling device enables the drilling of ice with a bit having a diameter up to 350 mm. The total drilling depth, including the snow cover, is 2 m. The rotational speed of the bit is 160-490 r.p.m. and bit feed 0-2 m/min.

Compare the ice-cutting speed in various designs of ice-drilling units:

Non-self-propelled Fershtut—Pashchenko ice-drilling unit ($D = 350$ mm; $n = 1100$ r.p.m.:

$$v = \pi Dn/1000 = (3.14 \times 350 \times 1000)/1000 = 1.2 \text{ m/min.}$$

Non-self-propelled model 26, Vyborg Plant ice-drilling unit ($D = 320$, $n_1 = 170$ r.p.m., $n_2 = 345$ r.p.m.):

$$v_1 = 0.17 \text{ m/min}; \quad v_2 = 0.34 \text{ m/min.}$$

Self-propelled ice-drilling unit ILB-2 ($D = 350$ mm, $n = 272$ r.p.m.):

$$v = 0.3 \text{ m/min.}$$

The ice-cutting speeds at various rotational speeds of the bit in the OLB-42 ice-drilling unit are 0.191, 0.32 and 0.52 m/min.

The slower the feed, the smaller the chips, and vice versa. The chips look like pulverized snow when ice is drilled with the Fershtut—Pashchenko unit, and are the size of hazelnuts when the ILB-2 unit is used. The power consumed in ice destruction is proportional to the volume of ice destroyed and also depends on the size of the chips. It is more advantageous to drill ice at high feeds (producing large chips), because the ice is not reduced in this case to the state of finely pulverized snow.

The cross-sectional area of the chips produced by various ice-drilling units varies within a very wide range. Thus, when drilling with the model 26 unit, $S = 3$ mm/rev, $D = 320$ mm.

$$f = SD/2 = 3.220/2 = 480 \text{ mm}^2.$$

When using the model OLB-42 ice drill, $f = 960 \text{ mm}^2$ at $S = 6$ mm/rev, and in the case of ILB-2 $f = 5250 \text{ mm}^2$ at $S = 30\text{-}35$ mm/rev, $D = 350$ mm.

When tough metal is drilled, flowoff chips are produced in the form of curls; non-flowoff breakoff chips are formed when brittle metal is drilled. Break off chips are characteristic of the ice-drilling process.

Data on the capacity of mechanical rotary drilling are presented in Table 1.4.

Table 1.4.

Design of drill	Drilling depth (m)	Hole diameter (cm)	Core diameter (cm)	Drilling speed (m/min)	Power consumed (kW)	Reference
(1)	(2)	(3)	(4)	(5)	(6)	(7)
Ice drill (Tomsk Fishing Trust)	0.9	32	-	0.6	2.6	[145]
Ice-drilling unit (Fershtut—Pashchenko)	0.4	30-35	-	0.3-0.6	4.4	[147]
Ice-drilling unit (LB-1)	0.6	35	-		4.4	[53]
Ice drill (N.E. Shlyaev)	1.4-1.5	40	-	1.5-3.0		[147]
Non-self-propelled ice drilling unit, model 26, (P.I. Pshenichnikov and I.N. Morozov)	1.25	32	-	0.5	3.3	[53]
Light ice drill (OLB-42)	0.6-0.8	35	-	0.2-0.5	3.7	[48, 148]
Tractor ice-drilling unit (ILB-1, ILB-2)	1.0	35		0.4	13.0; 29.0	[150, 151]
Self-propelled ice-drilling unit (PRAG-GPI-65)	2.0				54.0	[143]
Twist drill	1.0	38	-	96-234	-	[194]
Core drill	-	11	-	90-168	-	[194]
Tricone drilling bit (US Navy)	-	25	-	180	-	[194]
Electromechanical core drill	-	15.6	-	24	-	[243]

Iceland core drill	415	-	90	-	3.5	[243]
Electro-mechanical drill	100	14	10	100	-	[243]

1.4. Hydraulic Jet Methods

It has been determined experimentally that high-pressure continuous water streams may be used with success for the destruction of ice. This method is most effective in cutting ice sheets 0.5-0.6 m thick, and the cutting effectiveness rises greatly if the ice sheet is on a solid base rather than afloat.

When cutting ice, the specific dynamic pressure of the water stream should slightly exceed the compression strength of the ice, i.e. it should be at least 2.5 MPa for a coefficient of ice destruction of 0.5. A pumping unit developing a pressure of 30-40 MPa is necessary to maintain this pressure, assuming a realistic distance from the point of the nozzle to the bottom of the slot [187].

The speed of cutting thick, especially hummocky, ice by means of jets decreases greatly, while the power consumption increases [182]. A device was patented in 1971 for breaking up an ice sheet on a water body (Author's Certificate 442,108 USSR), and another was proposed in 1976 for clearing water areas of broken ice (Author's Certificate 719,914 USSR). The unit comprises a hydraulic jet assembly in the form of an ejector, containing an internal coaxial pneumatic pusher which houses a reciprocating piston and an explosive charge. The unit for clearing away broken ice is in the form of a pontoon, comprising a jet-like apparatus with a nozzle.

This method still evokes interest in several countries. The theme of hydraulic cutting of ice and frozen gravel has been studied at the department of Engineering Mechanics at the University of Alberta, Edmonton, Canada [214]. Ice engineering research has been carried out at CRREL (USA) on developing units for high-pressure jet cutting of ice on the walls and gates of river locks. The unit was developed on the basis of a self-propelled trailer with a high-pressure pump, having a capacity of 75 l/min at 390-590 MPa. The water jet is directed through a special nozzle at the ice-covered parts of the lock chambers [203]. CRREL supervised laboratory research of jet cutting of ice blocks at the University of Missouri at Rolla, and at the gas dynamics laboratory of the Canadian National Research Council. Field tests on the ice of Lake Michigan were carried out by the Illinois Technological Institute on contract to CRREL. The tests demonstrated that the power consumption exceeded economic limits and was three to four times greater than that required for circular saws, although the jet technology could ensure the required cutting capacity.

Quite a number of patents have been filed on the use of hydraulic jets for the destruction of ice.

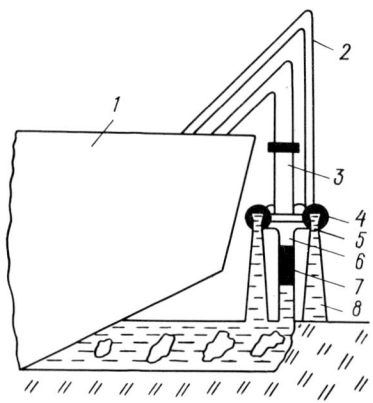

Fig. 1.74. Hydraulic jet ice-cutter [85].

1 - ship's bow; 2, 3 - low-pressure and high-pressure pipelines; 4 - toroidal unit; 5 - extension piece; 6 - nozzle; 7 - high-pressure jets; 8 - water anti-icing curtain.

Hydraulic icebreaker (pat. 3,977,345 USA) (Figure 1.74). A high-pressure conduit, ending with a nozzle, is fastened on the ship's bow. A toroidal unit with holes or extension pieces around the nozzle is connected with the low-pressure conduit. When the hydraulic icebreaker is operating, the jets form a water mist which causes icing-up of the ship's hull. In order to prevent this, water is supplied into the toroidal unit and a water curtain is created to eliminate spreading of the water mist.

Despite the great number of patents — for instance for propulsive nozzles to cut floating ice by high-pressure jets from beneath (pat. 3,877,407 USA), a hydraulic monitoring nozzle with a pneumatic lifting device (pat. 3,938,600 USA), and a hydraulic icebreaker (pat. 3,977,345 USA) — limitations of power consumtion have restricted the application of hydraulic technology to such spheres of ice engineering as cutting ice on runways and in locks, and laying pipes in ice and permafrost [230].

Hydraulic pulse destruction of rock is used widely in the extraction of mineral resources [115]. Ice destruction by pulse jets from a hydraulic giant (Author's Certificate 203,494 USSR) has been studied at the Hydraulics Laboratory of the Leningrad Institute of Water Transport and at the Ice Research Laboratory of the Arctic and Antarctic Research Institute.

IV-5 pulse water monitor designed at the Siberian Branch of the USSR Academy of Sciences. This disintegrated a 0.5×0.6×0.7 m ice block with one shot from a distance of 4.5 m, using only 1 l of water.

A passage was made in a 6.5×2.6×0.8 m ice block by 70 shots at the Ob water reservoir. The ice destruction capacity at a firing rate of 30 shots/min was 347 m^3/h, and the prime cost was 6 kopecks/m^3. The hourly capacity of breaking floating ice by the IV-9 water monitor was 6720 m^3 by volume and 10920 m^2 by area at a prime cost of about 0.1 kopecks/m^3, on increasing the nozzle diameter to 30 mm [183].

Table 1.5.
Effectiveness of ice destruction by IV-9 water monitor

Ice floe area (m^2)	Average ice thickness (m)	Ice floe volume (m^3)	Number of shots to break ice floe	Ice area broken by one shot (m^2)	Ice volume broken by one shot (m^3)
70	0.5	35	7	5.0	10
50	0.8	40	6	.6.6	8.3
35	0.6	41	4	5.3	8.9

Some data on the effectiveness of pulse destruction of ice are presented in Table 1.5.

The power ocnsumption is low, being 0.2-0.3 kW/h per 1 m^3 of ice on land and 0.01-0.15 kW/h per 1 m^3 of floating ice.

The pulse water monitor might be useful for breaking up ice on roads. American experiments have demonstrated, in particular, that the water clears away the ice from the roadbed without damaging it even at a pressure of 140000 kPa at small angles of attack [73].

An experimental prototype of an electric hydraulic monitor with a jet pressure of 3000 MPa has been designed at the Hydraulic Laboratory of the Leningrad Institute of Railway Transport [161]. Water monitors mounted aboard ships are illustrated in Figures 1.75 and 1.76.

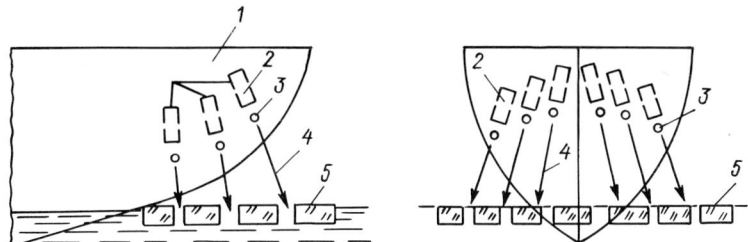

Fig. 1.75. Arrangement of water monitors in bow of ship for ice destruction from above [161].

1 - ship's bow; 2 - arrangement of pulse water monitors; 3 - holes in ship for water jets; 4 - hydraulic pulse jets; 5 - surface ice.

Three water monitors are mounted on each side in the bow of the ship to break ice sheets from above; one of the monitors is a standby, engaged only if necessary. The water monitors are arranged at various angles of inclination in the vertical plane. The angle and power of a shot are changed automatically, depending on the ice thickness (Figure 1.75).

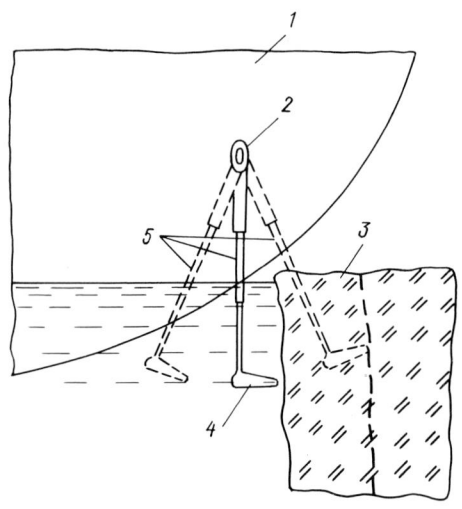

Fig. 1.76. Arrangement of water monitors for destruction of underwater ice [161].

1 - ship's bow; 2 - hatch in ship for outlet of water monitor with rod; 3 - thick ice; 4 - electric hydraulic pulse jets; 5 - surface ice.

An electric hydraulic monitor is mounted on each side of the bow of the vessel for underwater destruction of ice, in addition to the water monitors arranged for ice destruction from above (Figure 1.76). When the ice is being broken up, the electric water monitor is advanced to the underwater surface of the floe by means of a hinge and a special device arranged inside the ship. Vertical displacement is accomplished by extension of a telescopic rod to which the water monitor is fixed.

An icebreaker attachment with a pulse hydraulic gun has been tested in the USA. The device is a non-self-propelled barge with a powerful pulse hydraulic gun and a mechanism for ice removal. When the barge is pushed ahead, the ice broken by the pulse jets is lifted along an inclined guide and loaded onto longitudinal conveyers. These remove the pieces of ice to the upper parts of elongated slopes, and they then slide down under the influence of gravity beyond the edges of the channel. Taper and cylindrical rollers are arranged in front of the slopes to ensure uninterrupted movement of the ice [80].

1.5. Explosive Methods

An explosion is a process of rapid physical or chemical transformation of a system, associated with transition of its potential energy into mechanical work.

The work performed by an explosion is associated with the rapid expansion of gas or vapour, irrespective of whether they existed before the explosion or were formed during it. A sharp pressure jump in the immediate environment is the basic indicator of an explosion, and is the major cause of its destructive effect.

When the detonation wave reaches the surface of the explosive charge, the pressure at the wavefront begins to transform into a powerful pressure wave and into diverging motion of the environment (in this case ice, or water with ice). The pressure caused by the expanding gas products of the explosion drops rapidly. In the case of shattering explosives (for example, trinitrotoluene), the graph of pressure change shows a sharp peak attenuating gradually in an almost exponential manner. Attenuation is completed in a few milliseconds.

The disturbance from the explosion radiates in the environment in the form of compression waves. The wave has a sharp front in the nearest zone and a large amplitude; it is called a shock wave. A shock wave differs from a low-amplitude elastic wave in several specific features: (1) the speed of its propagation near the charge in water for example, is several times greater than the maximal speed of elastic waves in this environment; the speed of a shock wave decreases rapidly with motion down the velocity of sound; (2) the pressure amplitude in a spherical shock wave decreases faster than in low-amplitude waves, depending on the distance, (3) the profile of the wave expands gradually as it is propagated. All these features are most vivid in the high-pressure area, i.e. near the explosion.

A substantial amount of experimental data is available on the behaviour of solid bodies, including ice, at comparatively low dynamic and static loads, but the behaviour of solid bodies, especially ice, as a result of the high loads arising from explosions has been insufficiently studied. Unlike a fluid, which returns to its initial condition after the removal of any practically attainable load, ice is characterized by so-called residual deformation; beside that, its crystalline structure may change or even disappear at high pressures. Ice is in a plastic condition, and it is necessary to consider its compressibility under conditions of propagation of strong stress waves of loading (space wave) and unloading (reflected wave).

A series of studies, mainly of an experimental character, have been carried out on the problems of breaking a floating ice sheet by explosive means [77, 110-113, 232, 239]. Special charges of great fougasse capacity (ammonite, plastite, etc.) have been developed for underwater explosions. The authors of [18] used pulse cine radiography to study the specific features of ice deformation in the zone near the explosion. It was possible to estimate the role of internal friction in the process of destruction, and make some specific recommendations for engineering calculations on breaking an ice sheet by means of an explosion.

Experimental research has made it possible to determine the optimal depth for sinking charges to gain the maximal effect. When solving local problems (making channels in ice bridges or lanes in an ice sheet), a small amount of power is necessary, which is proportional to the volume and mass of the ice being broken. When an underwater directional explosion occurs near the bottom surface of an ice sheet, the shock wave reaches the ice—air interface almost without attenuation owing to the good acoustic contact of the water with the ice. It is then reflected, causing more intensive destruction than in the case of blasting a superimposed charge, or a charge first sunk into the ice sheet. In the second case, a substantial part of the shock wave energy is dissipated because it does not encounter the essential difference in the density of the media at the ice—water interface (this is certainly the case if the water layer is thick enough [17].

The method of a directional explosion was used by S.E. Nikolayev when cutting a channel in the shore-fast ice of Antarctica [110-113]. In the case of directional blasting, the major outburst of ice uccurs along the line of least resistance. The effect of destruction in this case is 1.5 times greater than in the case of a concentrated explosion. A double charge is used: the auxiliary charge explodes first, and the main one 0.025 s later (Figure 1.77). The gas cavity of the main explosion expands and rises to the ice surface, while the auxiliary explosion compresses and increases its diameter, flowing over the gas sphere of the auxiliary charge [113].

The method of blast-hole explosion of uneven ice surfaces with small charges has been used in the construction of runways; subsequent levelling and rolling of the surface with a tractor is necessary [112].

Directional blasting of sea ice sheets is carried out as follows.

When the charges are ready, electric detonators are inserted and the power supply circuit is laid, with a view to retarding the explosion of the main charge.

The main and auxiliary charges are sunk through holes under the ice and are arranged at a certain horizontal distance from one another. The auxiliary charge is usually placed in the water just at the bottom edge of the ice, while the main one is sunk to a depth which is determined by calculations.

Current is supplied when the charges are connected to the common power line. The auxiliary charge is the first to detonate. As a result of the effect of the shock wave, it takes 0.025 s for the compression of water, development and expansion of cracks in the ice, and displacement of the water and ice. The latter is accompanied by minor heaving of the ice surface at the location of the auxiliary charge. The gas bubble formed as a result of blasting the auxiliary charge reaches its greatest dimensions, and the pressure inside it is reduced to atmospheric. There is still no outburst of gas from the water to the air through the ice.

The main charge explodes about 0.025 s after the first. The gas pressure is transmitted to the surrounding fluid at the moment of the explosion, but it is not the same in all directions. The main pressure of the gas and water compressed by the explosion rushes toward the gas sphere formed by the explosion of the auxiliary charge. Some part of the explosive energy is consumed in the comminution and outburst of ice located directly over the main charge. The action of an underwater explosion, especially a directional one, depends greatly on the depth to which the main charge has been sunk. The ice breaks up and tears away from the massif under the shock-wave effect of the charges (both main and auxiliary).

As a result of the work done by the gas from the explosion of the main charge, the comminuted ice is ejected with water at a certain angle to a considerable distance from the site of the charges. About 60% of the ejected ice settles in the direction of the throw, about 15% is ejected to the sides and only about 10% is thrown to the rear of the explosion direction.

In the case of a directional explosion the zone of destruction has

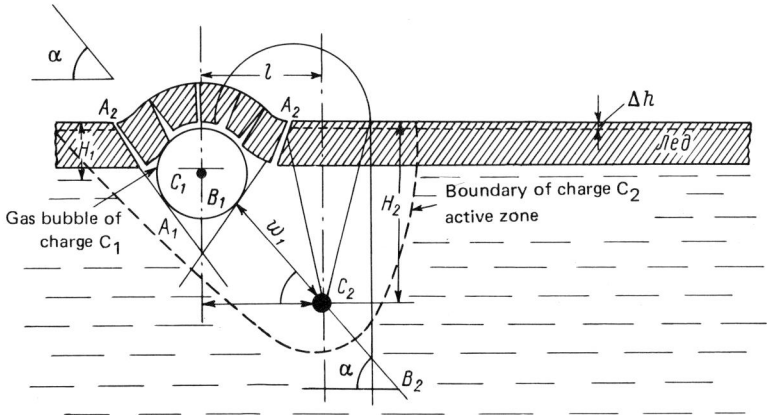

Fig. 1.77. Schematic of directional explosion of ice [113].

C_2, C_1 - main and auxiliary charges; H_2, H_1 - depth of main and auxiliary charges, w_1 - perpendicular distance from main charge to boundary of bubble formed by auxiliary charge; A_1-A_2 - tangent to gas bubble of auxiliary charge at contact with perpendicular; B_1-B_2 - perpendicular to boundary of bubble formed by explosion of auxiliary charge; 1 - horizontal spacing between main and auxiliary charges.
Indicated in the figure are, on the left-gas bubble of charge C_1, and on the right the boundary of the charge C_2 active zone.

the form of an ellipse, but when spot (concentrated) and muffling charges are applied it is circular.

Radial and circular cracks propagate in all directions from the zone of total destruction. The cracks are through ones, 5-10 cm wide, and propagate about 5-7 m from the edge; they then narrow down to hair width.

Results of experiments on breaking Antarctic sea ice by a directional explosion are presented in Table 1.6. The data demonstrate that the zones of total ice destruction by a directional explosion are almost always greater than the zones produced by means of concentrated charges of equal mass.

Directional blasting under optimal conditions ensures a zone of total destruction 50-70% greater than a concentrated charge. The outburst of ice on exploding a concentrated charge is less than 5%, while in the case of a directional explosion it reaches 50%.

The outburst of ice is greatly impeded by the force of ice cohesion with water, and the viscosity, incompressibility, and internal frictional force of the water.

The type of explosive is of great significance. There are two basic forms of external work done by an explosion: shattering and fougasse action.

Table 1.6.
Results of experiments on breaking Antarctic sea ice 1.1 m thick by directional explosion [113]

Charge mass (kg)		Charge depth (m)		Horizontal spacing between charges (m)	Delay time (s)	Outburst angle	Entire destruction zone diameter (m)	Destruction zone area (m)	Lane cleanness (%)
Main (C_2)	Auxiliary (C_1)	Main (H_2)	Auxiliary (H_1)						
25.2	3.6	3.0	1.4	2.4	0.025	33°	16.1 × 14.7	185.4	30
25.2	3.6	2.8	1.2	2.2	0.025	36°	16.2 × 14.6	185.6	35
25.2	3.6	2.8	1.2	2.0	0.025	38°	16.9 × 15.2	202.8	40
25.2	3.6	3.2	1.4	2.0	0.025	42°	17.9 × 15.7	220.4	45
25.2	3.2	3.2	1.2	2.0	0.025	45°	18.8 × 17.3	256.8	50
25.2	4.8	3.4	1.4	1.8	0.025	49°	16.9 × 13.2	176.2	30
25.2	4.8	3.4	1.4	1.5	0.025	53°	14.6 × 13.1	149.0	20

The shattering action causes comminution and deformation of the ice. This action is related to by the impact of the detonation products (shock wave) under very high pressure.

The fougasse action is displayed as splitting and scattering of the ice. This action is the result of the expansion of the detonation products at a relatively low pressure. The capacity of the detonation products (gases) to produce work during their expansion is called explosive strength or fougasse capacity.

Trinitrotoluene, characterized by a high velocity of detonation (6800 m/s) and relatively high density (1600 kg/m^3), mainly causes an impact on the ice, which is necessary for comminution but less so for the outburst. Hence, the use of trinitrotoluene is of low effectiveness in this respect. It is best to use an explosive with a high fougasse capacity, for example, alumite, plastite, ammotrinitrotoluene, etc., or to enlarge the main charge 1.5-fold to produce a clear lane by means of directional destruction of the ice.

The explosion of the main charge is delayed 0.025 s or 0.050 s by means of electric detonators with delay fuses DF No. 1 and 2.

Other time intervals are ensured by means of a detonating cord and EDDF No. 1 and 2, interconnected in various combinations (Table 1.7).

Table 1.7.
Methods of delaying the explosion of charges for destruction of sea ice [113]

Delay interval of main charge explosion (s)	Connection of main and auxiliary charges	Charge explosion frequency (s)		Explosion delay means		Notes
		Auxiliary	Main	Auxiliary	Main	
0.001	parallel	0.014	0.025	DC, 98 m	EDDF No.1	DC detonating velocity 7000 m/s
0.018	parallel	0.007	0.025	DC, 49 m	EDDF No.1	
0.025	series	instantaneous	0.025	ED No.8A	EDDF No.1	
0.030	parallel	0.020	0.050	DC, 140 m	EDDF No.2	
0.039	parallel	0.011	0.050	DC, 77 m	EDDF No.2	
0.050	series	instantaneous	0.050	ED No.8A	EDDF No.2	
0.057	series	instantaneous	0.057	ED No.8A	EDDF No.2 DC, 49 m	

DC - detonating cord; ED - electric detonator; EDDF - electric detonator with delay fuse.

Table 1.7 shows that in the case of a 4.8 kg auxiliary charge (all other conditions being equal) the optimal explosion delay time of a 25.2 kg main charge is 0.025 s, which ensures a maximal zone of entire destruction and a maximal outburst of ice. The explosion is considerably less effective if the delay time is increased or decreased (Figure 1.78).

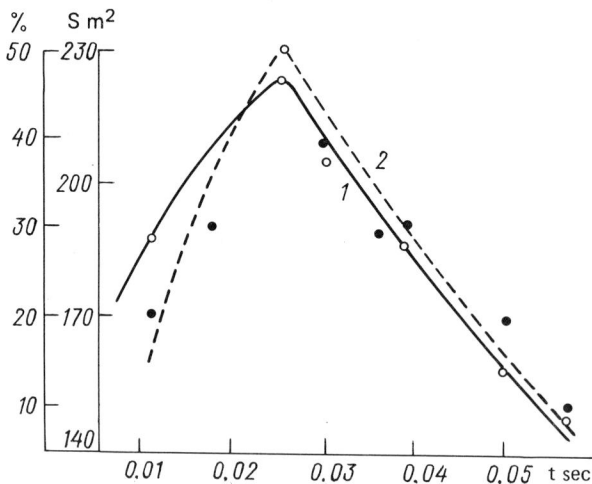

Fig. 1.78. Dependence of destruction area (1) and ice outburst (2) on delay time of main charge explosion [113].

The outburst angle is of great significance in directional destruction of ice, i.e. the angle of explosion direction α. If the angle is less than the optimal, the explosive energy may be insufficient to remove the ice to a certain distance. If the outburst angle α is greater than the optimal, a directional explosion is like the explosion of a double charge in character.

Table 1.6 demonstrates that the maximal zone of destruction and maximal outburst of ice are ensured at an explosion angle of 45°. For this angle the area of entire ice destruction is maximal and the outburst is 50%. The optimal explosion angle at ice destruction is also 45%, i.e. when the horizontal spacing between the main and auxiliary charges is the same as the difference in height. The outburst of ice is maximal at this explosion angle.

When blasting an ice sheet by means of a directional explosion, it is advisable that the mass of the auxiliary charge should be 1/6 that of the main one. The mass of the main charge depends entirely on the ice thickness and is calculated from the formula

$$C_2 = k \cdot H^3,$$

where C is charge mass, H is the ice thickness (m) and k is a coefficient depending on the properties of the explosive and ice.

Figure 1.79 illustrates the dependence of the destruction area and ice outburst on the mass of the auxiliary charge.

MECHANICAL DESTRUCTION OF ICE 99

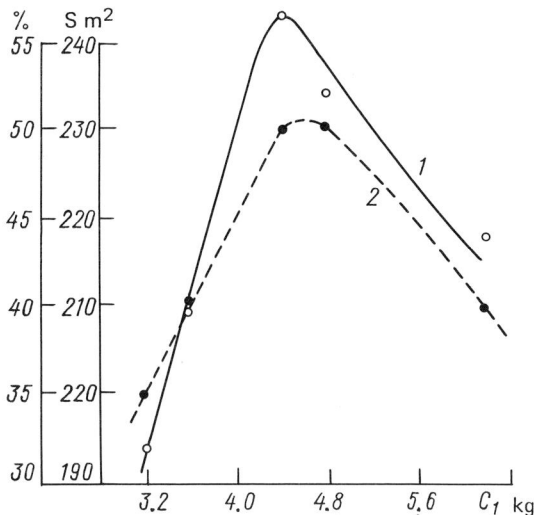

Fig. 1.79. Dependence of destruction area (1) and ice outburst (2) on mass of auxiliary charge (C_1) [113]. The mass of the main charge is 25.2 kg in all cases.

The sinking depth for the auxiliary charge depends greatly on its mass and is usually equal to the distance from the centre of the charge to the water surface. In making calculations, it is necessary to bear in mind that the maximal radius of the gas bubble, formed by the explosion of the auxiliary charge should be less than this distance.

The optimal sinking depth for the main charge depends on its mass and is found in specialist handbooks.

The line of least resistance of a directional explosion (at optimal parameters) is determined by the formula

$$\underline{W}_1 = \underline{l}\sqrt{2} - \underline{R},$$

where $\underline{l}$ is the horizontal spacing between main and auxiliary charges (m) and $\underline{R}$ is the maximal radius of the auxiliary charge gas bubble (m).

The thickness of the ice sheet in the calculations should include the snow cover.

When navigating in ice, the necessity for an explosion arises if it is impossible to overcome a heavy ice bridge separating the ship from open water. Figure 1.80 illustrates a scheme for blasting designed to break through such an ice bridge. Small charges seated in the ice in a line 80 m ahead of the ship's bow contributed to weakening the ice so that the icebreaker could make a passage itself. Then five charges were sunk under the ice in a line 220 m long, which made it possible to break the ice bridge [125].

Figure 1.81 illustrates a scheme for blasting heavy hummocky ice with a thickness up to 3.2 m.

Figure 1.82 illustrates a scheme for blasting to break off an ice beak, and Figure 1.83 shows a scheme for blasting ice beaks on floating ice fields. A blasting sheme to protect ships and evacuate them from the zone of compression is shown in Figure 1.84. The ice pressing on the

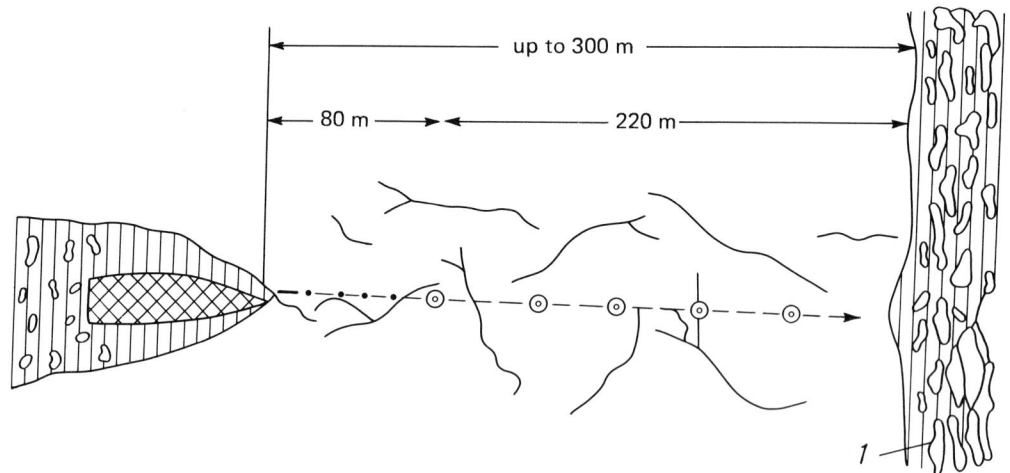

Fig. 1.80. Blasting scheme for breaking an ice bridge [125].

1 - ropacs. The area up to 300 m from the explosion is indicated.

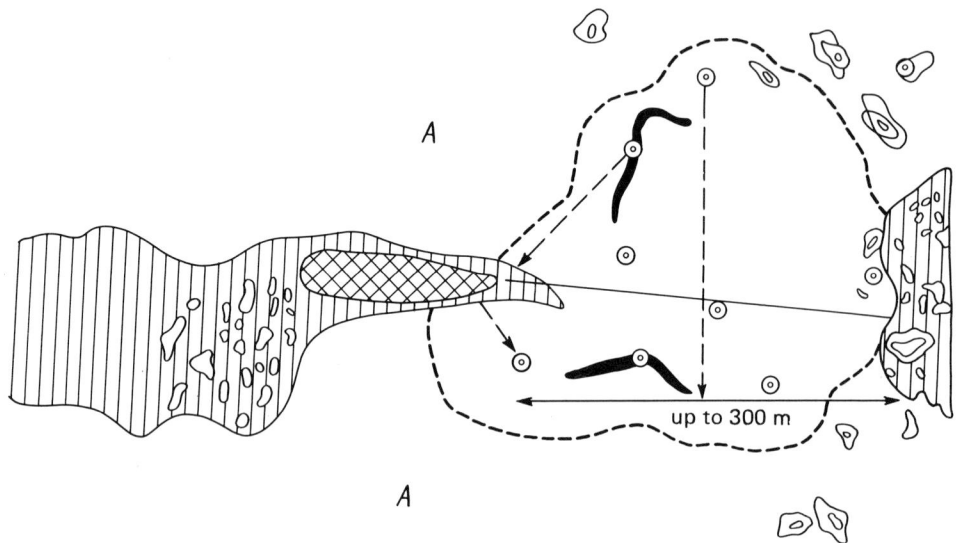

Fig. 1.81. Blasting scheme for heavy hummocky ice [125].

A - ice field.

MECHANICAL DESTRUCTION OF ICE   101

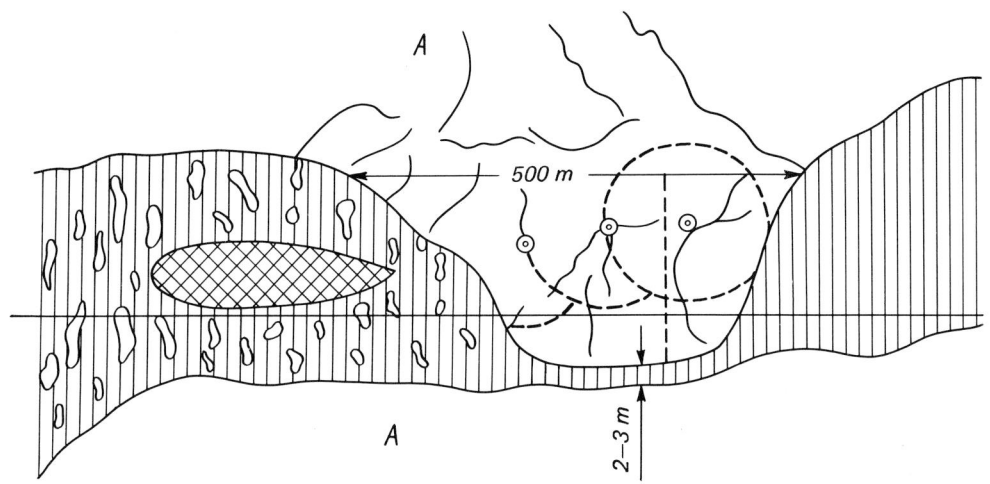

Fig. 1.82. Blasting scheme for an ice beak obstructing navigation [125].

A - ice field.

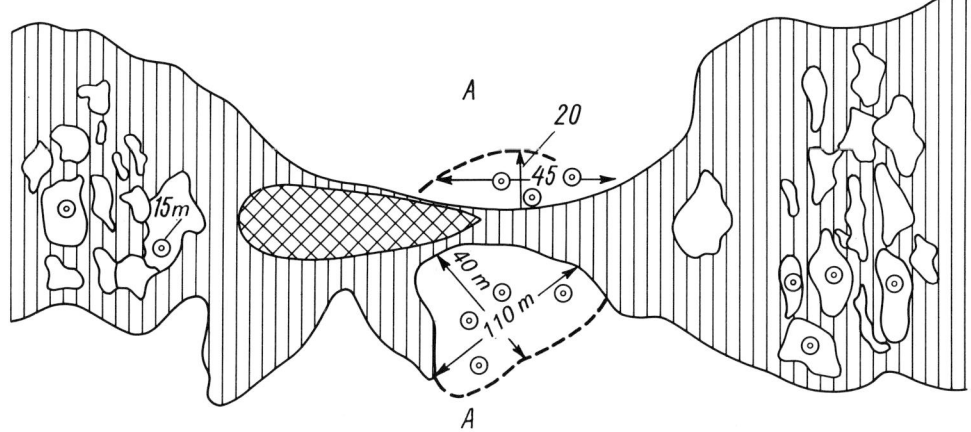

Fig. 1.83. Blasting scheme for widening a passage between two ice beaks [125].

A - ice field.

ship was blasted with small charges to starboard. The solid ice sheet was blasted portside with strong charges, creating a so-called 'ice-cushion', ensuring absorption and dispersal of the ice pressure on the hull.

As is shown in Figure 1.84, a large amount of explosive was needed to make a channel for the course of the ship, which is marked in the

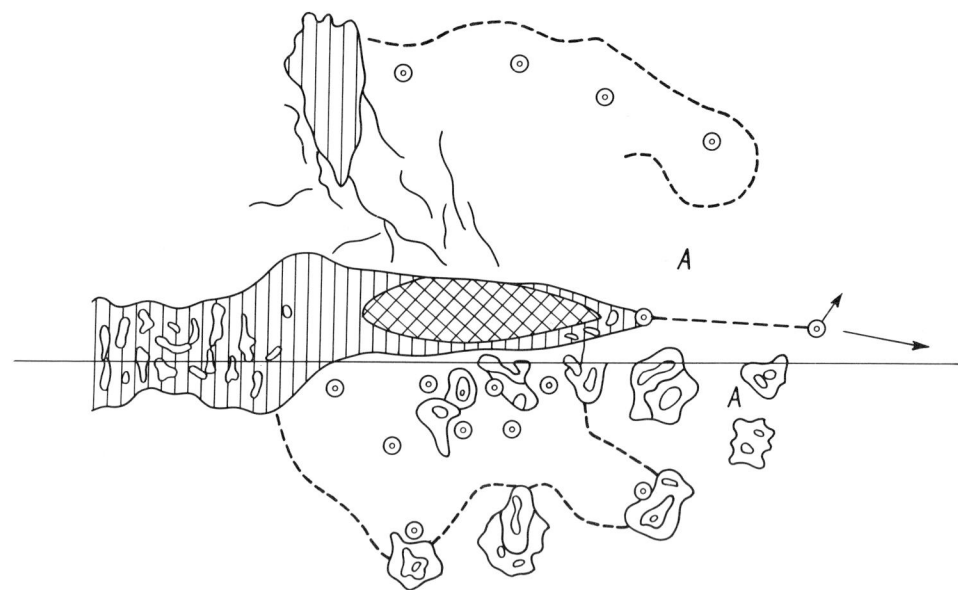

Fig. 1.84. Blasting scheme to relieve compression of a ship in heavy ice [125].

A- ice field.

figure by an arrow. The blasting operations made it possible to reduce the compression by creating an air cushion.

Blasting may be used on rivers to:
- carry out preventive measures to ensure free navigation in the ice in a certain part of the river;
- carry out preventive measures to protect bridges and water-development works against drift ice;
- eliminate ice jams at the time of their development, and the cause of their appearance (large fields, their initial accumulation, etc.);
- eliminate ice jams already formed;
- eliminate blocking drift ice.

Explosions cause great harm to the fishing industry. Because of environmental protection laws, the use of blasting to prevent jams (removal of ice from areas, creation of lanes, channels, etc.) should be considered inadvisable. Given permission from higher authorities these operations may be carried out only in exceptional cases, i.e. if the probability of ice jams in a particular year is great and the losses caused by them will exceed those of the fishing economy.

The application of blasting methods in eliminating river ice jams and blocking drift ice is discussed below in accordance with the Methodological Instructions [88].

Blasting operations should be started at the first indication of melting snow and rising water. The specific consumption of explosives is taken as 0.5 kg/m² when calculating the charge mass to loosen ice in flood-plain lakes and stagnant water bodies.

In some cases lanes are made to protect structures against drift ice when the ice sheet is very thick. The dimensions of the lane depend mainly on the width and thickness of the ice sheet, the length of its motion, the depth of the river near the banks, and resistance of the structure to the ice pressure.

Thus, the length of a lane in an average river downstream from a bridge should be not less than the width of the ice sheet, and twice this width upstream from the bridge. The total length of the lanes in small rivers should equal the width of the ice sheet. The length of the lane upstream of a bridge is increased in some cases to 500 m and more. The width of lanes at a bridge across a small river is taken to be equal to the width of the ice sheet.

Considering the specific consumption of labour required to sink the ice in the absence of a flange or a polynya, lanes are usually made by throwing the ice out on to the ice sheet by means of mass blasting of charges. A maximal outburst of ice is ensured by a single-row arrangement of the charges at a distance of ±.5-2.0 m from one another in the depth of the ice, and a specific consumption of explosives of 0.9-±.5 kg/m³. However, the width of the lane in this case is not usually more than ±5 m. Two-row and three-row arrangement of the charges is necessary to make wider lanes.

In the first case the charges are arranged opposite one another, and in the second case they are staggered. Underwater charges of the same mass are used for two-row blasting. If three-row blasting is used to ensure a clearer lane, the mass of the charges in the middle row is ±.5-2 times greater than the mass of the charges in the outer rows. All the charges are exploded simultaneously.

External and internal charges are used to make narrower lanes, as well as ice holes, and to break individual floes.

Charges sunk under the ice are exploded to clear large parts of rivers from ice to prevent ice jams. The explosions are made upstream, usually in series, so that the broken-off floes are carried downstream freely.

The mass of the underwater charge is calculated by the formula

$$Q = kH^3$$

where Q is the charge mass (kg), k is the specific consumption of explosive (kg/m³) and H is the depth of sinking a charge into the water (m).

The specific consumption of an explosive varies from 0.3 to 1.5 kg/m³, depending on the diameter of the lane, the required degree of ice disintegration, and its scattering.

At $\underline{k} = 0.3$ kg/m³, a lane will not form and the floes break up into individual pieces, while at $\underline{k} = 0.5$ kg/m³ a lane forms with a diameter which is 3.5 times greater than the sinking depth of the charge in the water. The lane's diameter exceeds by a factor of 4 the sinking depth of the charge at $\underline{k} = 0.9$ kg/m^3. The lane is quite clear of broken ice in the latter case. The cost of the operations is minimal when exploding charges at a depth of 1.5-3 m.

The charge sinking depth H increases with an increase in the ice thickness. The spacing between the charges depends on the diameter of the forming lane, blasting conditions, and the character of the operation. The spacing usually varies from 5 to 15 H (Table 1.8).

When splitting the ice or forming a lane next to an object, the spacing between the charges is 5 H and the spacing may be increased to 15 H in the case of flanges of polynyas.

It is not recommended to sink charges under the ice through cracks, pools, or joints between floes, because it is dangerous to approach them. A charge is usually sunk under the ice on a strong cord, one end of which is tied to the charge proper, and the other to a crosspiece arranged across the hole. The charge may be sunk on a pole. The charge is tied to one end of the pole, and the other end of the pole is fastened to the crosspiece. When sinking a charge, the Bickford fuse is brought to the surface if necessary. The fuse is tied to the twine or pole in several places. A Bickford fuse must not be used to sink a charge.

Table 1.8.
Mass of underwater charges and spacing between them for destruction of river ice of various thickness [88]

Ice thickness (m)	Charge sinking depth (m)	Spacing between charges (m)		Mass of charge (kg)	
		5 H	15 H	$k=0.5$ (kg/m^3)	$k=0.9$ (kg/m^3)
0.3 - 0.4	1.4	7.0	21.0	1.4	2.5
0.5 - 0.6	1.6	8.0	24.0	2.0	3.6
0.7 - 0.8	1.9	9.5	28.5	3.4	6.2
0.9 - 1.0	2.3	11.5	34.5	6.1	10.9
1.1 - 1.2	2.7	13.5	40.5	8.8	17.7
1.3 - 1.5	3.3	16.5	49.5	18.0	32.3

Operative interference into the process of ice jamming implies blasting preferably at upstream locations most unfavourable from the viewpoint of probable ice jams. The latter include:
- a leg of maximal ice thickness. The ice should be blasted at the bank and a little downstream;
- a sudden turn in the course of the river. The ice should be blasted at the banks within the length of the turn;
- a narrowing of the river-bed. The ice should be blasted at the banks along the narrow leg and within the subsequent widening;
- a leg with a shallow-reach alteration. The ice should be blasted near the banks at the reach leg downstream from the shallow;
- an island. The ice should be blasted at the shore and at the upstream and downstream ends of the island;
- a sand spit. The ice should be blasted at the banks and on the spit proper.

MECHANICAL DESTRUCTION OF ICE

Ice fields are blasted by underwater and external charges. It is impossible to approach large floes at the beginning of ice drift, when the ice drifts downstream in a solid mass. They are blasted in this case by charges that are thrust from shelters on the bank, from light bridges suspended on steel cables between steep banks, or from helicopters. Strong shelters are made to protect the blasting personnel against scattering ice fragments.

The mass of the thrust charge should not exceed 2 kg. A floe can be blasted more effectively, and with less danger for the blasters, by an underwater charge when the ice drift is not massive. The charges should be thrust into the water ahead of the floe so that the explosion occurs when the centre of the floe is just over the charge. If it is possible to approach the floe by a boat, or walk across the ice on boards or suitable wooden ladders, it may be blasted by sinking a charge into the water (Tables 1.9 and 1.10).

Table 1.9
Dependence of underwater charge mass on transverse size of floes (at ice thickness approx. 60 cm) [88]

Transverse size of floe (mm)	10-15	16-20	21-30	31-40	41-50	51-70
Underwater charge mass (kg)	0.5	1	2	4	6	12

Table 1.10
Dependence of external charge mass on river ice thickness [88]

Ice thickness (m)	0.2-0.3	0.4-0.5	0.6-0.7	0.8-0.9	1.0-1.2	1.3-1.5
External charge mass (kg)	1.2	2.0	3.0	4.5	7.0	10.0

The mass of underwater charges should be at least 15-20 kg to blast floes with a size in excess of 70-70 m. If the floes are sufficiently strong, they can be blasted by several underwater charges.

The charges are seated where the stream and wind contribute to the removal of blasted ice. The jam is eliminated by disintegrating it gradually in the direction opposite to the stream.

Ice frequently blocks small rivers down to the bottom and creates obstructions 3-4 m high under the water surface. These jams are eliminated by a series of strong blasts, the charge being arranged in the middle of the jam along the river stream.

If the jam is stable, subsequent charges are seated in the same places after preliminary removal of broken ice.

Charges weighing at least 30-40 kg are sunk into the water between the floes on rope loops to accelerate elimination of a strong jam.

The ice sheet should be blasted if a jam starts to form on a controlled part of the river by diving floes or by floes hummocking at the edge of an immobile ice sheet. The ice sheet on narrow rivers should be broken in the upstream direction, beginning from a spot where the jam is causing no damage.

If a jam has formed and stays in place despite elimination of the ice sheet downstream, the ice should be blasted at the head of the jam in several places along the bank or along the river channel.

If a jam is formed which thrusts towards the banks, it is necessary to blast the ice fields or the ice at the bottom edge of the jam, and if the jam nevertheless stays in place, a series of blasts should be made along the jam. When a jam is formed in a narrow river, it should be blasted along its length downstream, or simultaneously along its entire length. This helps to form a channel in the jam along which the water can flow from the upper to the lower reaches. The water level is gradually lowered and the jam is washed out.

As stated above, jam blasting is of low effectiveness in the narrow river if the length of the jam is more than 15-20 km. However, it may sometimes be useful to blast the ice along the entire length of the jam (or downstream).

When a jam forms in a wide river by floes diving or hummocking at the immobile edge, the ice sheet should be blasted downstream from the jam just as in the case of narrow rivers. If the jam stays in place after eliminating the ice sheet downstream and the latter becomes more compact, a series of blasts should be made along the bank, or in the middle of the river at the most compact part of the jam. It is advisable to make the blasts simultaneously. If the jam has already formed, then in many cases it is unnecessary to blast the ice sheet upstream. Successful blasting of the downstream, most compact, part of the jam usually causes a break in the jam. Hummocky accretion of floes, drifting downstream, disintegrates the ice sheet within the boundaries of a homogeneous part of the river. The accretion of ice usually stops at an obstruction and a new jam begins to form.

Ice should be blasted during the hours of most intensive solar radiation, i.e. from 12.00 to 15.00 local time, because intercrystalline ice layers melt at an above-zero temperature under the effect of radiation.

Charges are exploded in the initial period of ice blocking by thrusting charges from river banks, boats, iceboats, etc. These blasts contribute to uninterrupted passage of cream ice, which is frequently delayed at the banks. These spots should be inspected periodically to eliminate shore ice and to let delayed ice drift downstream.

Ice blocking in winter occurs more frequently under the ice cover and it can be eliminated at the time of its formation by blasting charges under the ice, sinking the latter through lanes. The charges should be arranged in longitudinal rows, so that a water passage can form immediately along the entire length of the ice block.

Ice blocks should be eliminated whenever possible because it is very difficult to break them after formation. It has been necessary in some cases to use 10 tons or more of explosives to blast an ice block, while in other cases they could not be blasted at all.

Blasting is currently successfully performed with the use of helicopters, which deliver the blasting teams and explosives, and also participate in the direct elimination of jams. Blasting with the aid of the MI-4 helicopter is carried out by landing the blasters on the ice, or directly from the helicopter. The helicopter hovers while the blasters disintegrate the ice.

The first method enables blasting of single and group charges. The blaster goes out on the ice, seats the charge or sinks it under the ice, returns to the helicopter, ignites the Bickford fuse, and throws it on the ice. The helicopter then moves over to the next charge. The length of the first Bickford fuse should ensure that its burning time enables the setting and ignition of several charges, and removal of the helicopter to a safe distance. The burning time of the first fuse is controlled by a stopwatch and a control ignition tube.

In the second case, when the helicopter hovers over the ice, the blaster on command lowers a charge with a connected detonating cord on a rope to the ice, or into the water between the floes. When the charge is seated, the blaster connects an ignition fuse to the end of the detonating cord, ignites it and lays it on the ice.

The operations are carried out in strict conformity with the 'Technical Rules for Blasting Operations' and 'Safety Precautions at Blasting Operations'.

The main advantage of helicopter operation is the ability to a charge almost anywhere. Compared to bombing, the use of helicopters offers greater accuracy in seating the charges and, consequently, greater effectiveness of blasting operations. Bombing is ineffective as a means for destroying ice over a large area: an increase in the power of the bomb increases the degree of destruction only insignificantly [59, 60].

Channels made in an ice jam by explosions, and blasting the ice along the jam, are effective measures [88].

The theory and practice of destroying icebergs by bombing and blasting [232], and by means of torpedoes with thermal heads is being developed in some countries. A torpedo is dropped on an iceberg by a helicopter and, melting the underlying ice, sinks into the iceberg by gravity. The exploding hexogen charge ensures maximal effect in disintegrating the iceberg [92].

The US Arctic Oil Association has been financing field tests on destroying ice islands by blasting since 1972. The results have been studied and then used in Antarctica to support operations carried out by the US Navy in 1975-1976. According to M. Mellor and A. Kovacs, destruction of a floating ice sheet by means of charges arranged directly beneath the bottom surface of the ice was most effective. The crater's radius exceeds the ice thickness by a factor of 8, and the effectiveness of ice destruction was characterized by the volume of disintegrated ice per unit of the charge mass, i.e. 7.8 m^3/kg.

The effectiveness is only 2.2 m^3/kg when blasting ice with shelves. The method of directional blasting is used to produce even surfaces in cutting ice islands, and it is necessary to seat a considerable quantity of explosives according to a predetermined scheme.

According to Western data, the cost of blasting a small iceberg or an ice island (volume up to 100000 m^3) is \$20000 [232].

Small-sized cumulative charges of the most effective explosives may be used to eliminate icing of underwater parts of ships and hydrotechnical structures.

Ice on the outside surface of plugs in sea openings of ships may be destroyed remotely by blasting elongated cumulative charges seated in advance when sealing the openings.

1.6. Inventions for Mechanical Destruction of Ice[1]

Nos	Country	IPC, NPC	Patent No.	Filed	Year of publication	Filed by
1	USSR	17B,1	134,275	4 Mar. 1960	1960	A.F. Nikolayev

For trenching in ice. Comprises vertically rotating pipes, having flanges on the surface with cutters, and a rotatable auger inside the pipes to eject comminuted ice. Openings in the pipe surface for discharging comminuted ice facilitate auger operation.

2	USSR	17B, 4/01	138,627	9 Dec. 1960	1961	A.F. Nikolayev, S.S. Solovyev

For cutting channels in ice. A self-propelled machine with a driving auger to remove snow from an ice surface, having a suspended cutter at the rear. To reduce power consumption in cutting a channel, the cutter is contained in a casing which is open on the bottom and side of the cutter member. The housing comprises a superimposed pipe supplying compressed air to remove ice chips from the cutter to the water.

3	USSR	17B 4/01	142,310	10 Apr. 1961	1961	A.F. Nikolayev, S.S. Solovyev, E.E. Rams

For trenching in ice with hollow pipe cutter to cut ice. To ensure maximal effect in removing chips, the cylindrical housing contains a driving mechanism with an air intake neck and a gate on the inoperative side of the cutter, forming a circular cavity in conjuction with the trench walls to catch and displace chips.

4	USSR	17B 4/01	143,816	28 Oct. 1960	1962	A.F. Nikolayev, I.F. Fedin, D.R. Butayev

For cutting slots and trenches in ice with side-milling cutter. To increase the capacity and reduce the specific power consumption, the cutter is hollow; it is in the form of two discs in a hub having a secured flange with a spur cylindrical gearing on the inside with two rows of cutters on the outside.

5	USSR	$65a^1$, 3	147,468	22 Feb. 1960	1962	I.S. Peschansky, Z.I. Shvaistein

Ice-cutting ship with an apparatus in the prow to cut ice into strips. To lift and eject the cut ice onto the ice field, the ship's prow is lowered below the bottom surface of the ice cover and the upper plane

[1] Patent information retrieval for 1960-1980 by M.N. Korovina.

of the prow is inclined with a gradient toward the ship. A chain conveyer is mounted to the rear of the inclined plane to lift ice chunks.

| 6 | USSR | 17B, 4/01; F25c | 195,473 | 22 Mar. 1965 | 1967 | A.F. Nikolayev, A.O. Vaganov |

For cutting slots in ice on water bodies. Comprises a self-propelled carriage and a driving cutting machine (pipe-milling cutter with an auger to remove ice chips). To simplify the design and make it lighter, the carriage drive is in the form of two cylinders with toothed working surfaces, one of which is fixed on a vertical shaft, joined kinematically with the cutting member drive.

| 7 | USSR | 17B, 4/01; F25c | 195,474 | 23 Mar. 1966 | 1967 | N.V. Cherepanov |

For drilling holes in ice on water bodies. A holder fixed on a rod with a detachable cutter. The holder is helical to centre the cutter in the hole and for convenient use, and comprises a means to adjust the ice chip thickness.

| 8 | USSR | 17B,4; | 217,408 | 16 Sept.1963 | 1968 | A.F. Nikolayev et al. |

Ice-cutting machine. Comprises a milling cutter in the form of a rotatable pipe with reversible transporting auger inside, and outside cutters arranged in a helical line. Holes are arranged in front of each cutter to cut complete and incomplete slots in the ice.

| 9 | USSR | 65a,3; B63B | 217,222 | 24 Apr. 1865 | 1968 | M.I. Kur |

Icebreaker with vibration unit inducing oscillatory motion in the elements of the ship's hull. To increase ice navigation of the ship, the vibration unit is a power cylinder with a freely moving mass which develops inertial forces in reciprocating motion, inducing oscillatory motion of the ship and dynamic shocks transmitted via the ship's bottom to the ice edge.

| 10 | USSR | 65a,3 | 227,117 | 13 June 1967 | 1968 | L.V. Ivanov |

For removing fine broken ice accumulating under the bottom of a ship navigating in ice. It is in the form of a scraper. To reduce power consumption in freeing a ship from ice, the scraper comprises two symmetrical elements forming an angle in the plane with the apex directed opposite to the ship's motion and coinciding with the diametral plane of the ship dragged over it.

| 11 | USSR | B63B,35 | 285,523 | 15 Sept.1969 | 1970 | V.M. Ivanov et al. |

For improving the cleanness of a channel astern of an icebreaker having a propeller propulsion system by sliding broken ice under the ice field. To simplify the process of ice removal, the propellers are turned 12-15° outwardly from the diametral plane.

| 12 | USSR | B63B,35 | 285,524 | 23 Jan. 1968 | 1970 | Moscow Institute of Transport Engineers |

For cutting lanes in ice cover by cutting the ice into cards and pushing the latter under the ice cover. Two buoyant machines are used.

| 13 | USSR | 17B,4; F25c,5 | 279,650 | 28 Apr. 1969 | 1970 | A.F. Nikolayev, A.A. Nikolayev |

For cutting slots in ice on water bodies. A cylindrical-milling cutter with cutters and apertures along a helical line. Comprises an internal auger to displace ice chips and a pipe for supplying compressed air for more effective removal of the chips from the internal cavity of the milling cutter.

| 14 | USSR | B63B; 65a,3 | 287,532$^+$ | 7 Aug. 1965 | 1970 | V.V. Rastorguyev et al. |

Pushed icebreaking and ice-clearing apparatus with a vibratory system. A hinged sliding lock is mounted on the rear wall of the apparatus across the diameter and on the prow of the pushboat. One part of the lock rotates around the coupling axis, the other is a vertical rail connected to the coupling.

| 15 | USSR | B63B,35 | 310,837$^+$ | 23 June 1969 | 1971 | G.Ya. Serbul |

Pushed icebreaking equipment. A pontoon with sledgelike bow and stern, the pontoon comprising ballast tanks to ensure effective icebreaking. Vertical runners are arranged on the bottom at the sides of the pontoon, and a cutter in the form of a vertical strip is fixed in the diametral plane.

| 16 | USSR | B63B,35 | 315,341 | 14 Aug. 1968 | 1971 | Alexbow Canada Ltd. |

Prow extremity of an icebreaker hull with a stem tumbling-in stern with side surfaces to deflect the broken ice. To reduce the power consumption required to break the ice cover, the stem comprises an ice-cutting strip arranged in the diametral plane of the icebreaker hull. The cutting strip is inclined 30° to the waterline plane.

| 17 | USSR | B64d; 15/06 | 213,588 | 19 Apr. 1965 | 1968 | I.A. Levin |

For removing ice from the skin surface of aircraft, for example. Elastic deformation is developed in the aircraft's skin by exciting mechanical oscillations as a result of pulses applied to the skin perpendicular to the surface. To increase the effectiveness of ice removal and reduce power consumption, pulses are applied to the skin that are at most 1/4 the period of the natural oscillations of the aircraft skin.

| 18 | USSR | B64d; 15/06 | 213,590 | 15 Nov. 1965 | 1968 | I.A. Levin et al. |

For removing ice from the skin of aircraft, for example, comprising a generator of mechanical pulses. The generator is an induction coil of eddy currents connected to an electric discharge power source for direct generation of deformation in the skin, with a consequent decrease in potential local damage.

| 19 | USSR | B63B,35 | 317,567 | 24 May 1970 | 1971 | E.N. Tsykin et al. |

For ice destruction. Comprises a body with an adjustable cutter for increased effectiveness. A boom with a plate for removing the broken ice is hinged to the rear part of the body.

| 20 | USSR | E02d,17; E21B,3; E21c,1 | 326,295$^+$ | 4 Nov. 1967 | 1972 | A.F. Nikolayev, A.N. Varnachev |

Mounted unit for ice drilling. Comprises a body, a drill with a helix and a rotation drive having a hollow shaft with a flange around the drill. To increase drilling depth and operational reliability, brackets are fixed rigidly on the flange of the drill rotation drive hollow shaft with turning rollers on axles, interacting with the drill helix.

| 21 | USSR | B63B,35 | 347,240 | 13 May 1971 | 1972 | E.N. Tsykin et al. |

Ice cutter comprising a cutter and a suspension. It is mounted in a ship's prow. The suspension is mounted on a bracket extending forward of the ship, and comprises a rod with a cutter and pickups connected to a mechanism for vertical displacement of the rod.

| 22 | USSR | B63c, 1/02 | 375,114 | 18 Jan. 1971 | 1972 | N.D. Andreyev |

For removing broken ice from beneath a ship's bottom. Comprises scrapers and a buoyant element. To increase operating range and efficiency, the scrapers are arranged along a helical line on the cylindrical surface of a cylinder and rotate with the latter by means of an engine located inside a cylinder which is joined by tie-rods to the dock crinoline and presses the scrapers to the bottom with invariable force owing to buoyancy.

| 23 | USSR | E02B, 15/02; F25c,5/02 | 371,302 | 14 Aug. 1970 | 1973 | V.I. Makeyenko, V.S. Maltsev, V.V. Yegorov |

For cutting and removing ice from a canal. Comprises a cutting member with a drive, frame-sledge, impact plate and guide. To increase operational efficiency, the impact plate is composed of hinged parts, one of which carries spring-loaded levers with bearing skis attached to the ends.

| 24 | USSR | F25c,5 | 385,147+ | 28 Dec. 1970 | 1973 | A.F. Nikolayev |

For cutting and removing ice. Comprises a milling cutter with a drive, whose cutters and apertures are arranged along a helical line. The spherical milling cutter comprises cutters arranged on helical strips, and the internal cavity of the milling cutter is connected to a suction pneumatic unit by a flexible hose.

| 25 | USSR | F25c,5 | 400,782 | 2 Aug. 1971 | 1973 | L.M. Grudin et al. |

For cutting ice on water bodies. Comprises a mobile frame, a petrol engine with a reduction gear, a circular saw, and a mechanism for adjusting the position of the saw. Comprises an additional lateral circular saw, which is fastened parallel to the main saw on the reduction gear outfit shaft.

| 26 | USSR | F25c,5 | 498,463 | 22 May 1974 | 1976 | I.F. Fedin |

For cutting slots in ice. Comprises a frame fastened on a self-propelled vehicle and two cutting members with a drive. To increase the capacity, the means has a wedge-hammer mounted on the frame with reciprocating motion to break the forming ice bar into separate pieces.

| 27 | USSR | F25c, 5/02 | 492,237+ | 2 Oct. 1964 | 1975 | N.V. Cherepanov, F.D. Sokolov |

For ice drilling. Comprises a flat ring with a detachable cutting tool. The ring has spacings for convenient removal of ice chips from the cutting tool. The cutting tool is arranged at one end of the ring so that its blade faces the other end.

| 28 | USSR | F25c,5 | 522,385 | 21 Jan. 1975 | 1976 | A.F. Nikolayev, A.I. Shkoda |

For cutting slots in ice and solid soils. Comprises a self-propelled means for displacing the working member, which is a side milling cutter. To increase capacity and reduce power consumption, it comprises a device in the form of a multi-disc cutter to make transverse cuts in the ice along the course of the working member.

| 29 | USSR | F25c,5 | 531,002 | 8 Apr. 1975 | 1976 | V.P. Savin |

Ice drill. Comprises a brace with an auger and two detachable cutting knives. Two blades are arranged in the bottom part of the auger and rotate around a horizontal axis. The detachable cutting knives are fixed on the blades.

| 30 | USSR | F25c,5 | 536,372[+] | 10 Oct. 1975 | 1976 | A.F. Nikolayev, A.A. Nikolayev |

For cutting and removing ice. Comprises a working member with cutters, a drive and a suction unit to suck away the ice chips. Comprises a casing, housing the working member, and a cutting chain. The working member comprises two augers with opposite winding, lateral disks, and a driving sprocket between the augers.

| 31 | USSR | F25c,5 | 557,242 | 22 Aug. 1975 | 1977 | A.A. Nikolayev |

For cutting ice. Comprises a mobile frame, a working member, a drive with a pulley, and a mechanism for adjusting the position of the working member. Comprises rigid posts with bifurcating rests, pressure and guide rollers in the form of a continuous bandsaw to fix the working member.

| 32 | USSR | B63B,35 | 562,182 | 22 Nov. 1972 | 1977 | Air Logistic Corporation |

For breaking an ice cover. Mounted in the prow of a ship, it comprises extending brackets with ice cutters. To save power, the ice cutters are in the form of milling cutters on driving shafts, turning on left, right, and central brackets, the latter turning along an arc with a chord equal to the distance between the ice cutters of the right- and left-hand brackets, exceeding the maximal width of the ship.

| 33 | USSR | E21c,19 | 592,976 | 4 July 1975 | 1978 | A.F. Nikolayev, V.N. Khudyakov |

For ice drilling. Comprises a drive and a cylindrical milling cutter with cutters and apertures on its surface along a helical line. To ensure reliable drilling, the cutters are hinged on the milling cutter body. The milling cutter contains a pipe with apertures with limited rotation around its axis. The pipe comprises bosses for interaction with the tail part of the cutters.

| 34 | USSR | E21B,3 | 606,993 | 8 July 1975 | 1978 | A.F. Nikolayev et al. |

Rotary feeding mechanism for ice-drilling apparatus. Comprises a rotation drive, including a hollow shaft with a flange; brackets are fixed on the latter with rollers for interaction with the drill helix, and a feed drive. To simplify the feed drive and increase the drilling depth, the drive comprises a bevel and a cylindrical gear, and an additional spring-loaded roller. The gears are mounted on the hollow shaft and are kinematically connected with the roller, which is fixed with the potentiality of interaction with the drill helix.

35	USSR	25c,5; E21c,23	610,992	3 Jan. 1977	1978	A.F. Nikolayev et al.	

For drilling holes in ice. Comprises a rod and a cutting head with cutters, an impeller with a support, mounted on a rod with axial displacement; the cutting head comprises a guiding hollow cylinder.

36	USSR	25c,5; E21c,19	641,249	3 Dec. 1975	1979	A.G. Medvedyev	

For making holes in ice. Comprises a cutting head fixed on a brace, cutter holders, and cutters, diametrally arranged crescent-shaped posts, and a tape helix. To increase the capacity, one of the posts is arranged inside the helix, and the other outside.

37	USSR	E02F,5	620,522	22 Feb. 1977	1978	F.F. Kirillov

For cutting slots in ice. Comprises a side milling cutter, cogs, and a drive with a driving sprocket mounted on a frame. To ensure better utilization of milling cutter diameter on cutting-in, the external surface of the disc comprises depressions, the cogs are arranged in the depressions and can interact with the driving sprocket arranged over the side milling cutter.

38	USSR	E02B,15	664,903	24 Nov. 1976	1979	B.Ya. Stazhevsky et al.

Ice-cutting machine. Comprises a self-propelled chassis with a cutting member in the form of a jib, a sledge with an incebreaker, and a blade. The icebreaker is a flexible cable mat tensioned between the cutting member and the sledge by means of rods.

39	USSR	B63B,35	654,485	1 Dec. 1977	1979	V.V. Klyuyev

Icebreaking and ice-cleaning means. Comprises a working member in the form of ice-cutting knives, pontoons with levers, and bearing rollers. Mounted on these are hinged parallelograms, and catches moving apart.

40	USSR	H02G, 7/16	666,601	12 Dec. 1977	1979	V.Ya. Sinelnikov, S.E. Vakulenko, I.V. Trach

For removing ice from aerial power lines. Comprises a mobile driven carriage mounted on the wire, with ice cutters and an engine with a power supply. To expand the operation zone, the power supply source is in the form of a throttling-type transformer.

41	USSR	E02B,15; F25c,5	694,593	22 Apr. 1977	1979	A.F. Nikolayev, V.N. Khudyakov

For cutting lanes and slots in ice on water bodies. Comprises a self-propelled carriage with a drive, hydraulic motors, slide valves with an electromagnetic drive for controlling the hydraulic motors, and a cutting working member. Comprises electric pickups on additional brackets, and additional brackets on the self-propelled carriage.

42 USSR   B63B;   814,807   28 June 1979 1981   B.V. Bogdanov,
             35/12   V.I. Lyubimov

For icebreaking and ice-cleaning. Comprises a body with lines of the underwater part and sledge-type extremities, a unit for mechanical ice cutting, and an ice-spreading wedge, comprising two beams mounted beneath the bottom in the middle part of the body. To' increase operational effectiveness, each beam comprises a turning drive with a driving member and is mounted such that it can turn around a vertical axis passing through the diametral plane of the body.

43 Great   B63B,35;   1,215,529 15 Aug. 1968 1970   Alexbow Canada Ltd.
    Britain B7A

Nautical icebreaking structures. A bow construction for a vessel for use in icebreaking, comprising two sloping surfaces deflecting ice, and an upstanding splitter blade for splitting the ice.

44 Great   B63B,35;   1,284,868 9 Sept. 1970 1971   Jean Fiorvanti
    Britain B7A   et al.

For cutting a channel in a layer of ice. An icebreaker ship comprises 10 vertically arranged helicoidal members fastened by journals to the bottom and the deck. The helicoidal members contribute to forward movement of the ship; they comprise knives that cut into the ice. The pieces of broken ice are moved upwards (or downwards if the helicoidal members are reversed) to a passage, whence they are evacuated to the ship's sides and forcibly ejected to a certain distance by means of conveyers.

45 Great   E02B,15;   1,490,642 7 Apr. 1975   1978   Sea-Log
    Britain B63B,35;   Corporation
             E1H

Ice cutter for drilling platform. A drilling platform for operation in Arctic waters comprising a work platform supported on a vertically extending cylindrical column. The column is within a rotating sleeve with readially projecting cutter arms.

46 Great   B02B,15;   1,490,877 7 Apr. 1974   1978   Sea-Log
    Britain E1H   Corporation

Ice-cutting apparatus for floating structures operating in ice-covered waters. A vertical shaft with several radial cutter arms rotates round a column supporting the platform of a structure. The cutter arms terminate in cutting edges at the outer ends of the structure.

47 Great   B64D,15; 1,331,698 23 Sept. 1970 1973   B.F. Goodrich
    Britain   B7W   Company

Pneumatic de-icer: an elongated inflatable de-icer boot for an aircraft.
It comprises a series of passages, having a longitudinally extending
edge with a bleeder channel within the boot forming the rear edge of
the boot on the wing. The bleeder channel comprises several elements;
for example, distensible fibres between opposing faces of the channel
to ensure a flow of the gaseous inflating medium.

48 Canada   114-11   942,591   22 Sept.1971 1974   Sun Oil Company

For cutting ice in the path of a ship. Two cuts are made in the ice
parallel to the course of the ship. The cuts are made by a power saw
or high-pressure nozzles utilizing fluid under pressure.

49 Canada   255-65   955,581   1 Mar. 1972   1974   Wilmot G. Hill

Auger for drilling a hole in ice. The auger includes an elongated
support shaft provided with a handle mechanism at its upper end and
an auger bit rigidly secured to the lower end of the shaft.

50 Canada   255-64   978,180   28 Apr. 1971 1975   A.B. Arjohn

An ice drill with adjustable cutting edges. Comprises a way of setting
the required cutting angle. It can be used effectively for cutting
different types of ice.

51 USA   114-40   3,130,701 15 Aug. 1961 1964   P.O. Langballe

An icebreaker comprising a submersible vessel including an expansible
and inflatable bottom and a reinforced shell with a fixed icebreaking
cam at its upper surface, a source of a compressed gaseous medium
within the vessel, means for inflating the expansible bottom by the
compressed gaseous medium to modify its volume, and means for propulsion
of the vessel by exhausting the compressed gaseous medium from the
expansible bottom during deflation.

52 USA   B63B,35; 3,590   15 Aug. 1967 1970   Alexbow Canada
              114-41   Ltd.

Ship's bow construction. A bow for use in both icebreaking and normal
sea-going service takes the form of a bulbous bow projecting forward
below the waterline, the cross-section of the bow being relatively flat,
i.e. generally having a greater horizontal than vertical dimension.

53 USA   B63B,35; 3,521,592 13 May 1968   1970   M.W. Rosner
              114-42

Ice channel cutter. The cutter is a marine vessel whose prow is fitted
with several rotary vertically extending ice-engaging units each

presenting an array of radially extending ice-chopping blades or cutters. The ice choppers are rotated simultaneously with forward movement of the vessel to chop the ice into relatively small chunks and carry them sideways and rearwards, or downwards to allow passage of the vessel.

| 54 | USA | B63B,35; 114-40 | 3,670,681 | 15 May 1970 | 1972 | T.B. Upchurch |

Ship-mounted icebreaking system. A ship with a combination of a saw-toothed, upwardly-biased, under-ice stressing member and an above-ice, downwardly-projecting, variable-position, chisel-type ice-cracking system.

| 55 | USA | B63B,35; 114-42 | 3,678,873 | 2 Oct. 1970 | 1972 | Sun Oil Company |

For cutting ice. Removes ice encroaching on an offshore platform from an ice floe or ice located in the path of a vessel, by making a pair of cuts in the ice parallel to the direction of travel of the vessel or ice floe. The cut ice section being removed is broken away from the remaining ice mass by forward movement of the vessel or pressure of the ice floe against the offshore platform. An additional cut can be made between and parallel to the pair of parallel cuts to make two ice sections, in order to facilitate forcing aside of the severed ice. These cuts can be made by mechanical saws or by nozzles utilizing high-pressure fluids.

| 56 | USA | E02B,17; 61-46 | 3,696,624 | 2 Oct. 1970 | 1972 | Sun Oil Company |

Bucket wheel ice cutter. Bucket wheels are mounted on offshore platforms or ships' prows. The bucket wheels rotate in a generally horizontal plane and are paired in opposite directions so that no torque is imposed on the structure or ship. Multiple sets of bucket wheels can be used to cut a thick section of ice. The bucket wheels can be inclined to cut a larger vertical section.

| 57 | USA | B63B,35; 114-40 | 3,690,281 | 21 Dec. 1970 | 1972 | Esso Research and Engineering Company |

Stern construction for icebreaking vessels. As the vessel backs into its own broken channel, it will effectively shovel the ice pieces up and to the sides of the vessel, thereby clearing the channel to facilitate backing and simultaneously keeping the ice away from the vessel's screw.

| 58 | USA | B63B,35; 114-40 | 3,698,340 | 2 Mar. 1971 | 1972 | J.C. Wagner |

Icebreaking system for ships. A system comprising a ship-borne compressed-air-operated ram for positioning beneath the water. The ram breaks sea ice by powered upward impacts with the under surface of the ice, accompanied by water-hammer at rapid ram oscillation.

59	USA	B63B,35; 114-42	3,768,428	24 Nov. 1971	1971	Air Logistics Corporation

A sweep ice cutter. It comprises port and starboard ice-chipping cutters, and a traversing ice chipper or cutter. The ice chippers or cutters are mounted ahead of the vessel by means of brackets. Port and starboard ice chippers are positioned at points which extend at least to the port and starboard extremities of the vessel or structure, and preferably further so that chipped ice, etc. may readily flow along the side of the vessel or marine structure. On motion of the vessel the port and starboard chipping cutters cut the borders of a lane, and the central cutter cuts the internal ice field. The central cutter is in the form of paired milling cutters rotating in opposite direction. Each cutter is provided with an electric motor.

60	USA	B63B,35; 114-40	3,754,523	19 Nov. 1971	1973	Esso Research and Engineering Company

Icebreaking tanker. It has an elongated hull with a bow and stern and a catamaran arranged above and connected to the hull, the elongated hull moving substantially under the icepack while the twin hulls of the catamaran are arranged to cut through the icepack.

61	USA	B63B,35; 114-42	3,791,328	19 June 1972	1974	Air Logistic Corporation

Ice-removal track. The system consists essentially of a centrally located ice-cutter mounted on a vessel, two pivoting ice-cutters which traverse a path between the central ice-cutter and the extremities of an expanded framework mounted on the port and starbord sides of the vessel. The ice is cut into a series of triangular-shaped blocks by cooperative action of the transversing ice-cutters.

62	USA	E02B,15; B63B,35; 61-1	3,807,180	29 May 1973	1974	R.W. Worthing

Ice rifter. An icebreaking apparatus including a floating platform with buoyancy control means such as ballast tanks, valves, and a source of water connected to it. The platform includes apparatus for perforating or weakening the ice, such as explosive charges to provide perforations.

63	USA	B63B,35; 114-40	3,572,273	6 Aug. 1969	1971	Southwest Research Institute

For breaking a layer of ice on a body of water by repetitive combustive explosions. An apparatus for breaking the ice on bodies of water to allow passage of vessels by providing a movable buoyant body having a face for contact with the ice and several exhaust openings in the face and positioned below the waterline of the body. Each opening is connected to a combustion chamber for applying the combustion energy from a hydrocarbon fuel directly to the ice in order to break and melt it. Several exhaust openings in the hull of an icebreaker next to the bow, with a normally closed combustion chamber connected to each opening.

64 USA   B63B,35;   3,808,997 10 Oct. 1972 1974   Global Marine Inc.
            114-40

For clearing a path through ice. An icebreaking arrangement in which air is applied under sufficient pressure to lower the surface of the water at an ice-water interface to a level below the bottom of the ice over an area which extends on either side of the interface. The unsupported weight of the ice in the region in which the water level is depressed produces failure and breaking off of pieces of ice at the interface.

65 USA   B63B,35;   3,841,252 2 Oct. 1970 1974   Sun Oil Company
            114-40

For breaking ice. Two mounds of ice can be created by injecting gas beneath ice sheets at two locations to create a trough between them which is in stress and will therefore fail more easily on contact with a vessel or marine structure.

66 USA   B63B,35;   4,029,035 13 Apr. 1976 1978   W.H. German
            114-67A

Ship's hull and method of bubbling hot gas from it. A ship's hull having sides and several small holes arranged in one or more rows located well below the ship's waterline and extending along the sides. An apparatus for generating and supplying compressed gas to the holes at a pressure sufficient to produce continuous gas bubbling from the holes for the purpose of assisting in icebreaking.

67 USA   B62B,35;   3,913,511 9 Aug. 1974 1975   C.W. Weiland
            114-42

Navigational icebreaker. Two saw mounts are mounted on the bow portion of the deck. Each saw mount comprises a power means with a drive to shafts to rotate the saw blades. The rotating saws cut the ice in conjunction with forward movement of the apparatus. The port and starboard sides of the deck provide a structure for connecting the bow portion of the deck to a vessel. The saw blades score ice ribbons of substantially uniform width.

| 68 | USA | E21,5; 175-18 | 3,929,196 | 16 Aug. 1974 | 1976 | Uuno Raantanen |

Ice auger. Comprises an elongated tube with a helix to remove ice; attached to one end is a blade, and to the other a hand crank. The body tube consists of two sections connected by a hinge. One half of the hinge is affixed to the lower section, the other to the upper section of the body tube.

| 69 | USA | B63B,35; 114-42 | 3,965,835 | 29 Mar. 1974 | 1976 | Sun Oil Company |

Icebreaking ship. An ice-cutter is mounted on the ship's prow. Comprises a hollow cylinder arranged transversely to the direction of the channel and having teeth. The cylinder is mounted on a framework and can rotate around its longitudinal axis. The framework contains a power source. Thermal energy is discharged to the atmosphere through holes in the framework, creating an air-thermal screen near the cylinder and retarding freezing together of broken ice in the channel.

| 70 | USA | B63B,35; 114-42 | 4,005,665 | 23 June 1975 | 1977 | Sea-Log Corporation |

Fluid vacuum release for ice-cutting systems. A comminuting ice cutter comprising several rotating cutter elements, each having several cutting edges spaced circumferentially around a common axis and adapted to engage ice and dislodge pieces of ice by impacting action; a drive rotating the cutter elements in a direction around the common axis to apply high-velocity impacts on the adjacent ice; and means for directing fluid under pressure at the interface between each of the cutting edges of the cutter elements and the ice when it comes in contact with the cutting edges. This last dissipates by fluid ejection the partial vacuum resulting from ice cleavage and changes it into a positive pressure which aids the separation and removal of the chips by the cutter elements.

| 71 | USA | E21B,3; 173-26 | 4,057,114 | 7 Jan. 1976 | 1978 | Paul J. Anderson |

Hand-held ice auger. The auger is powered by a remotely located engine and comprises a flexible shaft with a coupling fastened to one end to attach it to a rotating shaft on the engine. A housing with an opening receives the other end of the flexible shaft. This housing contains a reduction gear, including a shaft which rotates at a substantially reduced rate.

| 72 | USA | B63B,35; 114-42 | 4,069,783 | 11 Nov. 1976 | 1978 | Sun Oil Company |

For disaggregating particulate matter. The destruction system comprises a rotatable drum with arrays of teeth extending radially from it.

MECHANICAL DESTRUCTION OF ICE

73 USA B63B,35; 4,070,062 11 Nov. 1976 1978 Sun Oil Company
         299-10

For disaggregating particulate matter. Parallel cuts with ridges between them are made by a toothed revolving unit. The ridges are then struck by a second toothed revolving unit.

74 USA B63B,35; 4,198,917 9 Aug. 1977 1980 Mitsui Engineering and Shipbuilding Company, Limited
         114-42

An icebreaking device for a ship comprising at least two icebreaking units each having a screw body in the form of a cone, provided with a peripheral spiral blade. These units are positioned at and extending beyond the periphery of the bow portion of the ship with the axes of the screw bodies lying in a vertical plane, the axes being perpendicular to and offset from the central axis line of the ship. The screw bodies can be raised and lowered on the respective sides of the central axis line. The icebreaker forms grooves in a mass of ice by the rotation of the screw bodies. The icebreaking units can be driven independently of each other in the same or opposite directions.

75 USA E02B,17; 4,230,423 23 Nov. 1977 1980 Mitsui Engineering and Shipbuilding Company, Limited
         405-211

Icebreaking apparatus for structure used in icy waters. A marine structure is protected against ice floes by an apparatus having a downwardly expanding frustoconical shape with several cylindrical rotatably mounted bodies. When the ice floe contacts at least one spiral blade, the latter is driven into an edge of the ice floe and either raises or lowers the ice floe to subject it to a bending force. This causes the ice floe to break into relatively large plate-like pieces which are pushed sideways by rotary action of the cylinders to effectively prevent damage to the central mounting by the ice floe.

76 France B64D,15 1,091,965 16 Feb. 1971 1972 Goodyear Tire and Rubber Company

Anti-icing device, particularly for aircraft. Comprises an inflatable skin of rubberized fabric, inflatable pipes and a header, connecting all the pipes to a common compressed air source, or to a suction device.

77 FRG B63B,35 2,206,472 11 Feb. 1872 1974 Frits Kallipke

Shape of ice-breaking ship with ice-breaking prow and external covering depressed at ice strake. The covering is depressed mainly at the ice strake and bends downwards from it to form a relatively wide plane gradually sloping outwards. The external covering slightly above the ice strake comprises upwardly sloping convex surfaces.

78	FRG	B63B,35	2,341,932	20 Aug. 1973	1976	Heinrich Waas

Icebreaker ship. Comprises a cutter fastened on a mobile carriage to cut slots in ice. At least one planing tool with an acute angle facing the ice surface is mounted on a carriage.

79	Japan	B63B,35; 84A0	46-33901	14 Aug. 1968	1971	Alexbow Canada Ltd.

Prow icebreaker. Comprises boards to deflect the ice, arranged on the port side and starboard side of the ship's prow. A central buttock forms at the mating lines of the boards' front edges. The buttock extends forward, gradually lowering to the main line of the apparatus, which is provided with a bearing structure below the waterline. The bearing structure is slid directly under the ice sheet. A cutting effect arises during motion owing to the skewed shape of the buttock.

Chapter 2

THERMAL DESTRUCTION OF ICE

Methods based on the thermal effect differ from others used for the destruction of ice cover. Comparing the expenditure for the destruction of a unit volume of ice by different methods, it is possible to show that the exposure of ice to the effect of heat to achieve its destruction is not the most advantageous method from the viewpoint of energy consumption. For example, under ideal conditions, when the total power is consumed for ice melting, a source of approximately 0.1 MW is required for complete melting of sea ice at a rate of 1 m^3/h. Nevertheless, thermal methods based on cheap energy, for example the waste heat of industrial or marine plants, or nuclear sources, may be convenient and economically advantageous, and in certain cases even the only way of overcoming problems associated with ice. Naturally, it makes sense to use the enormous free energy of the Sun to melt ice.

Calculations indicate (Figure 2.1) that 8.8 m^3 of ice can be warmed 18° (from -20° to -2°C) in 1 min with a 7.4 MW power source. Consequently, it is possible to warm about 530 m^3/h with the same power source, or an area of 530 m^2 at an ice thickness of 1 m.

When ice is warmed, its strength correspondingly diminishes (Table 2.1).

Subsequent warming of the ice by the Sun causes its destruction.

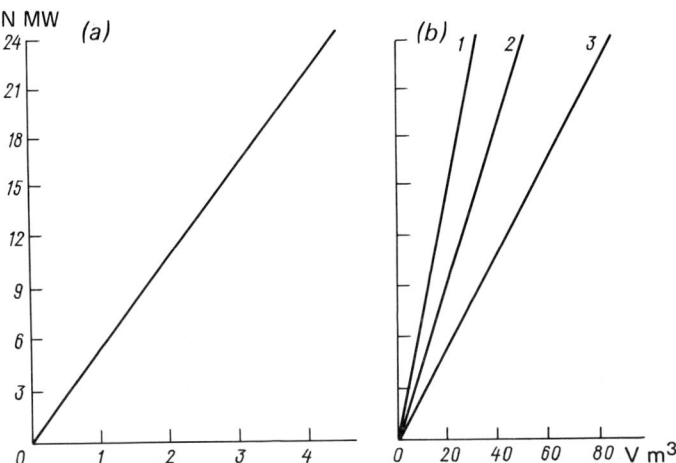

Fig. 2.1. Ice volume (V m^3) (a) melted or (b) heated in 1 min at various power consumptions (N MW) [175].

1 - from -20 to 2°C; 2 - from -10 to -2°C; 3 - from -5 to -2°C.

Table 2.1
Dependence of strength of one-year ice on temperature

Temperature (°C)	-20	-10	-5	-2
Ultimate bending strength (MPa)	1.6	1.2	0.8	0.4

2.1. Blackening

The solar radiation method, requiring no power consumption, is widely used in the USSR and elsewhere to accelerate breakup of ice sheets and melting of glaciers. This method is designed to ensure maximal use of solar energy for melting ice by reducing its reflectivity. A natural snow-covered ice surface is characterized by extremely high reflectivity. About 90% of solar radiation energy reaching this surface is reflected to the atmosphere. If a snow-covered or ice surface is covered with pulverized dark material, the reflectivity (albedo) decreases greatly, and more of the thermal energy of the Sun is used for melting. The blackening effect consists of violating the ice structure by penetration of blackening materials. [5, 9, 23, 24, 42-44, 56, 68, 139, 143, 154, 175-177, 198].

Dark-coloured powders with a density greater than that of water are used for blackening: coal and slag dust, phosphorite meal (mineral fertilizer), moulding sand (foundry waste), black sand, sand mixed with slag or coal dust at a ratio of 50% slag or coal. Phosphorite meal and slag dust are equivalent to coal dust in their effect on ice, so their application greatly reduces the expenditure on blackening.

Soot and black pigment (waste products of the paint and varnish industry) should not be used because of their carcinogenic properties. In addition, soot has no destructive effect on ice, because soot flakes aggregate into grainy clots that float on the water surface without penetrating into the ice.

The average consumption of blackening materials per square metre is determined on the assumption that the ice is totally covered with blackening materials to a height equivalent to one diameter of the blackening particles. This is calculated by the formula $n_{max} = \gamma d$, where γ is the density of the blackening material (g/m^3) and d is the diameter of the blackening material particles (m). The diameter should not exceed 0.5 mm, on the assumption that the blackening material should penetrate into the ice.

The maximal norms of consumption of blackening materials at $\gamma = 1$ t/m^3 are as follows: $n_{max} = 100$ g/m^2 (1 t/ha) at d = 0.1 mm; $n_{max} = 500$ g/m^2 (5 t/ha) at d = 0.5 mm. Material consumption and the cost of blackening operations rise sharply with an increase in diameter of the blackening particles. However, the reflectivity of the snow-ice cover dusted with blackening materials at amounts of n_{max} and 50% n_{max} differs only insignificantly. The consumption of blackening materials can therefore be reduced by 50% from n_{max} without much decrease in the

THERMAL DESTRUCTION OF ICE

effect of blackening. It is advisable to repeat the blackening operations after snowfalls when the fresh snow layer is more than 5 cm thick, but amounts of blackening materials may be reduced in this case to 15-20% n_{max} [88].

The graph in Figure 2.2 determines the amount of blackening particles per square centimetre of ice cover, depending on the diameter of the particles and consumption norms for a material of unit density. When the density exceeds unity, the consumption norms must be multiplied by the density of the material. The amount of blackening particles per unit of surface area decreases with an increase in the diameter of the particles (at the same rate of consumption of blackening materials), and the degree of destruction of the ice cover as a result of penetration of the blackening materials into the ice is therefore reduced likewise. Blackening materials of greater diameter are used for spraying from a high altitude (20-50 m and more), causing dispersal of the blackening material over a wide area beyond the planned route of blackening. (For example, river stretches with high banks and other projecting objects may prevent descent of the aeroplane down to 5-10 m.)

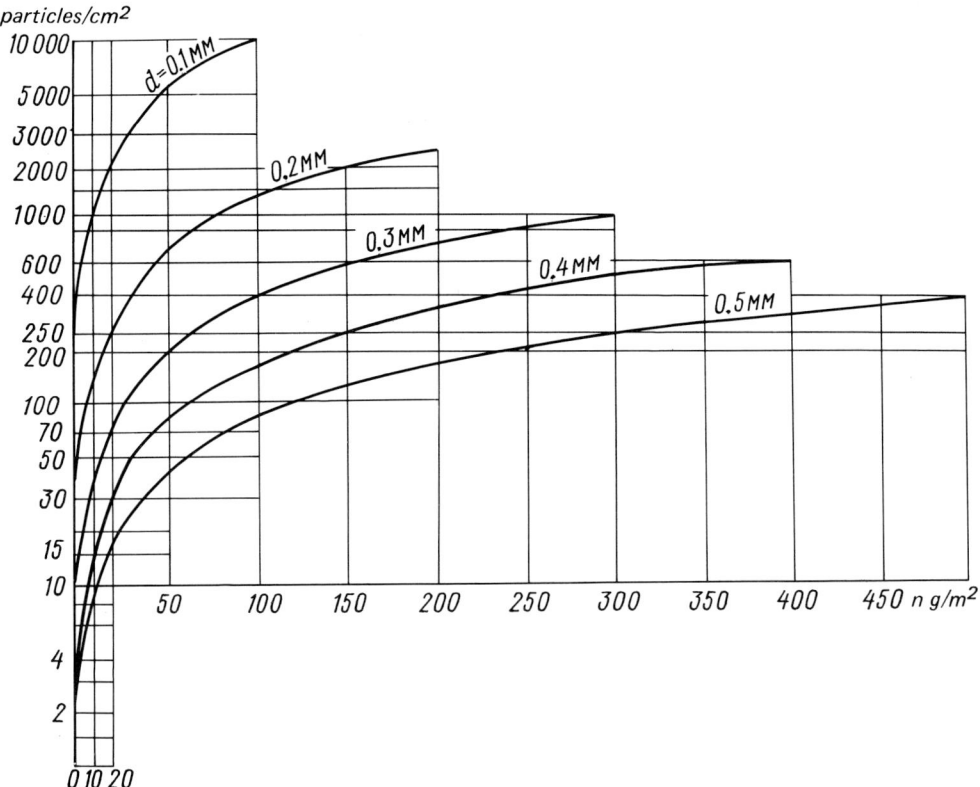

Fig. 2.2. Logarithmic curves of dependence of amount of particles per 1 cm^2 of ice cover on consumption norm (n) and diameter of dusted particles (d) [88].

The time to begin blackening operations depends on the ambient temperature and amount of snowfall. Blackening should be done after the end of heavy snowfalls in spring at an ambient temperature which eliminates top freezing as a result of melting blackened snow freezing on the ice.

The recommended data for blackening operations (neglecting the time required for the operations themselves) in regions with a continental or severe continental climate is after the time of stable transition of the diurnal positive air temperatures above the zero mark in spring; in regions with unsteady weather, characterized by frequent temperature falls (the mean daily temperature being above zero) and rises of the air temperature, operations can be undertaken after the time of stable transition of the mean diurnal positive air temperatures above zero.

Snowfalls after these dates have hardly any adverse effect on blackening, owing to rapid melting of the snow (1-2 days) as a result of solar radiation at above-zero air temperatures.

The dates should be established by consultation with the local weather bureau. If the area is affected by a stable anticyclone, blackening is possible before the air temperature rises above zero.

It is advisable to blacken the snow-ice cover at later dates, when it is melting rapidly and water accumulates on its surface. At this stage intensively melted snow and ice surfaces that have not been blackened will have almost the same absorptivity as blackened snow and ice.

The width of the routes to be blackened depends on the type of equipment used. When spraying from an aeroplane of the YAK-12 type, the width of the route may be 2.5-3.5 m, and, if an aeroplane of the AN-2 type is used, it is 8-10 m. The cost of spraying ice cover with blackening materials at various rates of consumption is 1.0 rouble/ha at a consumption of 100 kg/ha; 3.7 roubles/ha at 500 kg/ha; 7.2 roubles/ha at 1,000 kg/ha; and 23.0 roubles/ha at 3,000 kg/ha.

The blackening effect is increased by greater intensity and duration of solar radiation, as well as by the longer duration and higher temperature of the warm period preceding the break-up [24, 175].

The effectiveness of blackening on water bodies, sea bays and gulfs is several times greater than on rivers, where the period of maximal effective action of the blackening operation is reduced by the break-up of the ice cover by flood waves. The effectiveness of ice blackening can be estimated by a coefficient relating the absorptivity of blackened and non-blackened snow or ice:

$$k = (1 - A)/(1 - a)$$

where A and a are the reflection coefficients of blackened and non-blackened snow or ice respectively.

The effectiveness of top melting of blackened ice, when the snow melts away, can also be estimated to a first approximation by the coefficient η, indicating the relationship of the melting duration $\underline{T}$ of blackened ice to the melting duration $\underline{t}$ of non-blackened ice:

$$\eta = T/t$$

The Methodological Instructions [44] assist in the determination of other practical problems, e.g.:

- the minimal air temperature at which the ice begins to melt in the blackened layer, and in the case of a clean surface;
- the optimal thickness of the blackening cover;
- the optimal dates for blackening the ice;
- the consumption of blackening materials.

Spraying from an aircraft makes it possible to treat great areas over rivers in a short time [5, 143]. The ice albedo drops to zero and it breaks up two weeks earlier at a spray concentration of 50-100 g/m^2. The amounts of material required are slightly larger in Arctic regions: 300-400 g/m^2 of a coal dust and slag mixture. The most favourable dates are the beginning of May in the western USSR and the middle of April in the eastern USSR. The width of a strip blackened from a sledge hitched to a tractor is 7-10 m, and from an aircract 15-17 m [175].

When blackening ice in March on the roads of eastern Siberia with a 0.5-1.0 mm coal dust layer, the thickness of the melted blackened ice was 55 cm by 1 April, while the non-blackened ice melted only 25 cm [23, 24, 56]. Blackened ice melted about two weeks earlier than non-blackened ice on the roads of Tien Shan [23, 124, 186].

The experience of using the radiation method to weaken the effect of ice on a hydrotechnical structure has indicated that the destruction of blackened ice occurs 2-4 times faster than that of natural ice. The method of radiation weakening may be ineffective under conditions of frequent temperature falls, and it is enhanced by disintegrating the ice cover by furrows with the aid of an ice plough [177].

Artificial control of the melting of glaciers depends on their natural pollution, and can be increased by an order of magnitude by artificial blackening with coal slag. The use of this method on the glaciers of the Pamirs and Elbrus reduced the albedo 2-2.5 times and accelerated melting 7-11 times [9]. Ice blackening has been recognized as the most effective method on glaciers [75, 186].

A review of experiments on artificially enhancing ice melting by the radiation method has confirmed its effectiveness and economic profitability in destroying sea, lake, and river ice with a view to making navigable channels and extending navigation dates, as well as eliminating ice jams and blocking [42, 43, 88, 139].

Research in other countries, inspired by its success in the USSR, has indicated the great effectiveness of this method in extending the dates of navigation at the Canadian—American Isaacsen station in 1960 [195], and between Barrow and Prydz Bay in 1975 [198].

Research is continuing to discover substances that would reduce the ice albedo to the maximum. One of these substances (pat. 49-34582 Japan) is designed for spraying to enhance artificial icemelting.

2.2. Pneumatic and Hydrodynamic methods

Engineering studies have been carried out mainly in two directions in the search for new hydrothermodynamic methods of removing ice cover: use of the heat of deep waters by means of bubblers and flow generators, and artificial heating of a water body.

The idea of the first method is to supply compressed air through perforated pipelines or individual nozzles to the bottom water layers by means of a pneumatic (bubbling) unit. The buoyant air bubbles rise to the surface, carrying with them a mass of warmer bottom water. This, on giving up its heat, prevents ice formation or contributes to the melting of ice. This method is sometimes called <u>water body aeration</u>. The problem has been solved more successfully in some cases by the use of flow generators, whose basic advantage over bubbling units is in the better use of the heat rejected by the bottom water, because the flow generator creates a current along the lane which is in contact with the ice cover for a long time. If the heat reserve in the bottom layers is not great, it is possible to use artificial heating, i.e. specially preheated water.

Pneumatic units are currently used in the USSR and elsewhere for river transport, and in other branches of the national economy: in hydraulic power engineering (to create lanes upstream of hydrotechnical structures) and for timber rafting (to create ice-free water bodies to store timber).

Comprehensive laboratory and field research on the effectiveness of pneumatic units has been carried out at the Leningrad Institute of Water Transport and at the Central Research Institute of Timber Rafting. The accumulated experience is discussed in [10-12, 20, 51, 128, 133], where the use of bubbling units in freezing water areas with natural reserves of heat in the water is analysed. References [11, 51, 125] are the most interesting because they discuss the experience of making lanes when there is little or no available heat in the water. The results indicate that the bubbling units are effective at air rates characteristic of pneumatic breakwaters, when it is possible to break up ice at least 10 cm thick and create surface currents of at least 0.6 m/s. The air rate in these units is calculated by formulas derived as a result of field testing of pneumatic breakwaters.

E. Pounder, who considers this method to be effective only in the case of deep freshwater bodies under moderate climatic conditions, presents a comprehensive review of experiments on water bubbling in [122]. Detailed technical data on equipment for the destruction of ice and protection against ice formation are given in [50, 191], whence data have been adopted on the design of pneumatic units and flow generators.

The discharge of compressed air into the water through a bubbling device is common to all the pneumatic units. Two types of equipment have been developed: one supplies only compressed air into the water body, the other discharges compressed air and a heating medium separately or in a mixture. The specific features of the winter ice-thermal regime of the water body in question ought to be considered to ensure effective use of pneumatic units in both cases.

Assuming a knowledge of the thermohaline stratification of the water body, the compressed air is discharged into the layer of maximal vertical gradient of these parameters. An upward current forms owing to the 'piston effect'. Semi-closed circulation currents arise while the pneumatic unit operates (Figure 2.3), which can be divided into three basic areas: surface horizontal currents; vertical upward water—air currents; return currents that go back to the outlet of compressed

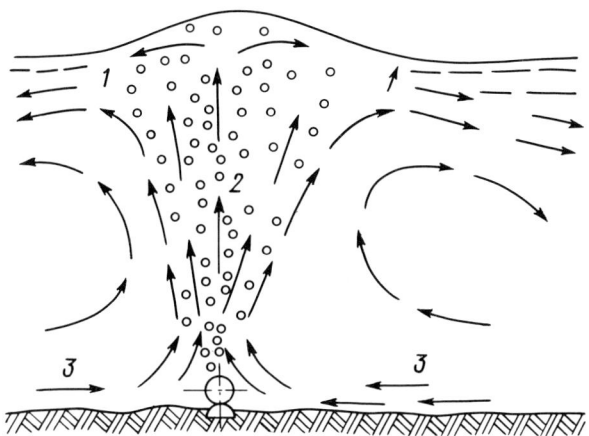

Fig. 2.3. Circulation currents created by bubbling unit [50].

1 - surface currents; 2 - vertical upward water-air current; 3 - return currents.

air.

Some idea of the formation, direction and rate of these currents is very important in choosing the dimensions and type of bubbler of a pneumatic unit under specific hydrological and morphological conditions.

It is important to arrange the bubbler in such a manner that all of the warm water should be involved in the circulation current during operation of the unit. In the case of sea water it is most advantageous to place the bubbler in the layer of maximal salinity, because a continuous flow of highly saline water to the bottom surface of the ice cover is necessary to melt the ice.

In the case of pneumatic units operated with the aim of utilizing the heat of waste industrial water, the bubbler should be arranged to ensure mixing of the incoming warm water with the bulk of the water body.

A vertical water-air current, in which a mass of water is lifted by air bubbles, has the shape of a conical flare. The upper expanded end of the flare, near the water surface, is transformed into horizontal currents. A feather is formed at the outlet of the vertical flow, creating a hydraulic head. The cooled water is replaced in the area of the surface currents by warmer or saline masses of water lifted by the air bubbles and interacting with the ice cover. The direction of the surface currents depends on the quantity of devices and outlet holes, and their arrangement with regard to the vertical walls of dockside structures.

In the case of a single spot source the vertical upward water-air current creates a fanlike stream on the surface, i.e. the current spreads radially from the centre. Several sources arranged in a row create a flat stream directed to both sides of this line in opposite directions and perpendicular to it. Only half of a flat stream, directed sideways, forms when the bubbler is operating near the wall.

The rate of the surface currents is reduced on moving away from the discharge of compressed air. At a given distance from the air discharge the rate depends on the sinking depth of the pneumatic unit bubbler and the specific rate of the compressed air.

The surface current reaches the maximal rate at a distance (0.3... 1.0) H from the spot of air discharge to the surface of the water body (where H is the sinking depth of the bubbler). When the bubbler operates near the wall, the rate of the surface current increases by a factor of approximately 1.3. These data are very important in calculating the zone of active effect of the pneumatic units, as well as the width of the channel along the pier which is kept free of broken ice owing to removal of the latter by the surface currents.

Pneumatic units can have a continuous or a pulse discharge of compressed air; these are the two major trends in their design and development.

In the case of pneumatic units with a continuous discharge of compressed air, the bubbler is a perforated pipeline. Such a device, resting on the seabed, with air supplied by a compressor on the shore, has been patented in several countries (e.g. USA, Sweden). The air discharged through the perforations into the water creates circulation of the warm bottom water to the surface of the water body. The pipelines may rest on the bottom in the horizontal plane in the form of sinusoidal or straight branches to increase the effective zone of the device. A device has also been proposed whereby the system of perforated pipes may surface. The pipeline is open at one end, while the other is fitted with a system supplying compressed air to displace the water. A valve is used to control the loading and emptying of the pipeline. The diameter of the pipeline is chosen so that the latter can surface together with all the devices and accessories mounted on it, when it is charged with air. The difficulty of orienting the device on the bottom is one of its disadvantages, but this can be eliminated by a device in the form of a double pipeline (Fig. 2.4).

The upper pipeline is fitted with nozzles to discharge compressed air, and is connected by clamps and cables to the bottom anchor pipe and weights. Stationary floats are fixed on the cables connecting the nozzle pipe to the anchor pipe. The floats hold the pipeline with the nozzles over the anchor pipe. Both pipes are connected to a compressor. The bottom pipeline is also fitted with a valve to fill it with water.

The anchor pipeline is charged with compressed air during installation, and equalizes the mass of the entire system; the valve of the anchor pipeline is then opened, it is filled with water and the system sinks to the bottom. The valve makes it possible to wash the nozzle pipeline with water, charge the anchor pipeline with air, and surface the system.

The trunk pipeline is fitted with an insulating jacket to prevent the formation of an ice plug.

As well as units using a continuous discharge of compressed air, pulse discharge units can also be found. Air is supplied to a bubbler with a special storage vessel; it is discharged into the water on

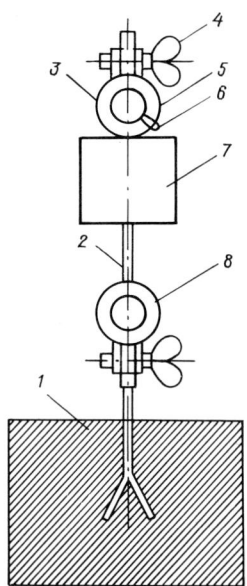

Fig. 2.4. Surfacing bubbling system [50].

1 - anchor; 2 - cable; 3 - perforated pipe; 4 - clamping screw; 5 - clamp; 6 - discharge nozzle; 7 - float; 8 - anchor pipe.

accumulation of a critical volume.

The pulse method ensures the discharge of air into the water in the form of large bubbles, and their surfacing rate increases with volume. The bubble takes the shape of an inverted saucer with a large horizontal axis, causing an increase in the amount of water carried upwards with each individual bubble. Another advantage of the pulse method is the fact that a rise of the bubbler capacity by increasing the air feed rate does not interfere with the bubbling regime of the stream in the vertical upward current, which may occur with a continuous-action bubbler when the air surfaces through the entire water thickness without ever mixing with it. The amount of water lifted by the bubbles and the operational effectiveness of the unit drop sharply in this case.

Bubblers with pulsed air supply are used mainly as breakwaters, but they are also effective for ice melting.

The operational effectiveness of pulse bubblers is also explained by the fact that a large air bubble brings to the surface along with it a large dome-shaped volume of water with a considerable hydraulic gradient, which ultimately ensures an increase of the initial velocity and operating range of the surface current created. This advantage has inspired the development of various designs of bubblers with pulse discharge of compressed air. One of the first prototypes of a pulse pneumatic unit was developed by a British firm and patented in Great Britain, France, the German Federal Republic, and other countries.

A pneumatic system with hydraulic pulse valve-storage (pat. 805,789 GB, 1,075,507 FRG, and 1,179,893 France) comprises an inverted container and two pipes arranged concentrically in the centre to make a waterlock (Figure 2.5). Compressed air is supplied through a pipeline to displace the water and fill the reservoir. On reaching a certain

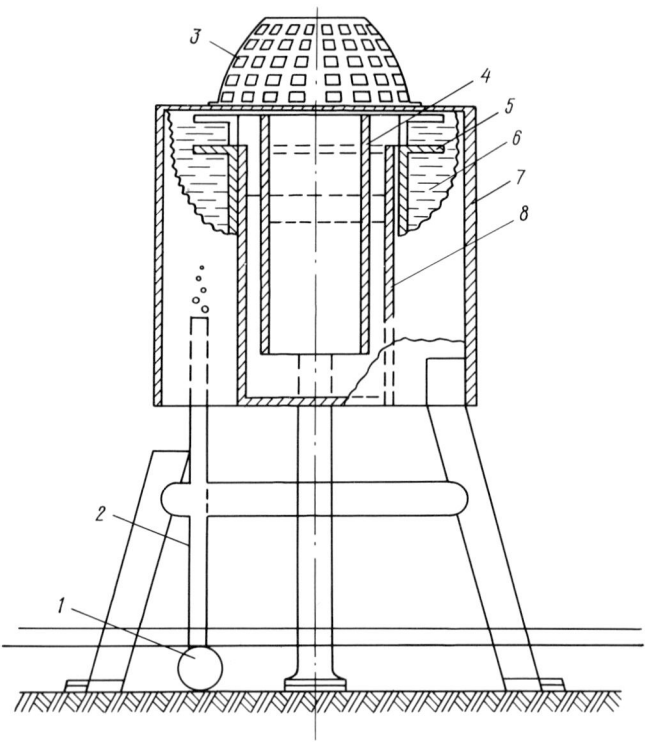

Fig. 2.5. Pulse valve-air storage [50].

1 - air line; 2 - air outlet pipe; 3 - screen; 4 - internal pipe; 5 - ring valve; 6 - screen; 7 - external pipe; 8 - body.

level the air instantaneously bursts into a central exhaust pipe, where it is broken up by a safety screen into individual bubbles and creates a powerful upwards air–water current.

Improved devices have been patented in Great Britain and the USA (pat. 967,543; 148,509 GB; 2,193,260 USA), comprising a valve and a vertical guide pipe (Figure 2.6). When tested in Canada, a pulse bubbler of the given type demonstrated high effectiveness at a low air consumption. A round lane with an area of 740 m^2 formed in a 0.3 m thick ice sheet after two weeks of operation at an ambient temperature of -23°C. The bubbler intermittently discharged one powerful air bubble with a volume of about 0.016 m^3.

The applicability and effectiveness of this device can be **judged**

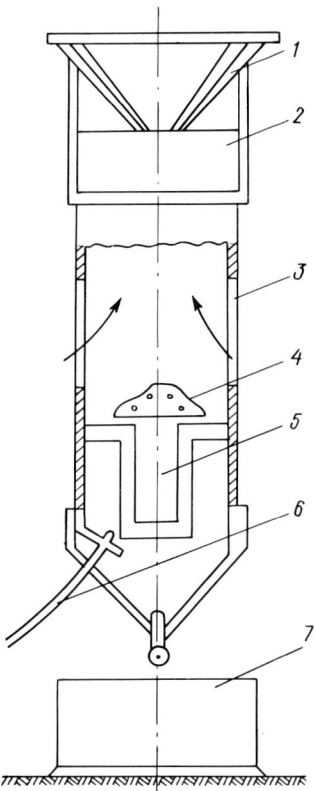

Fig. 2.6. Pneumatic hydraulic valve for pulse outlet of air [50].

1 - conical dissector; 2 - pipe; 3 - water holes; 4 - screen; 5 - valve-storage; 6 - air line; 7 - weight.

by the availability of analogous patents for the design of this device in several countries.

An apparatus for pulse supply of air has been developed in the USSR (Authors' Certificates 127,191; 127,192; 127,193) to melt ice cover on water bodies, and the design and technological characteristics of the elements of the units have been determined.

The disadvantage of pulse bubblers is that the air accumulating in the circular spacing of the waterlock discharges through the pipe in the form of a jet. The bubbles take shape in the water body, which limits the size of the bubble because disintegration, re-integration and further disintegration of the bubbles occur. Energy is expended in these transformations, and increased turbulence is generated in the zone of the vertical upward current.

The effectiveness of bubbling can be raised considerably by simultaneously warming the water. It is advisable to use equipment with

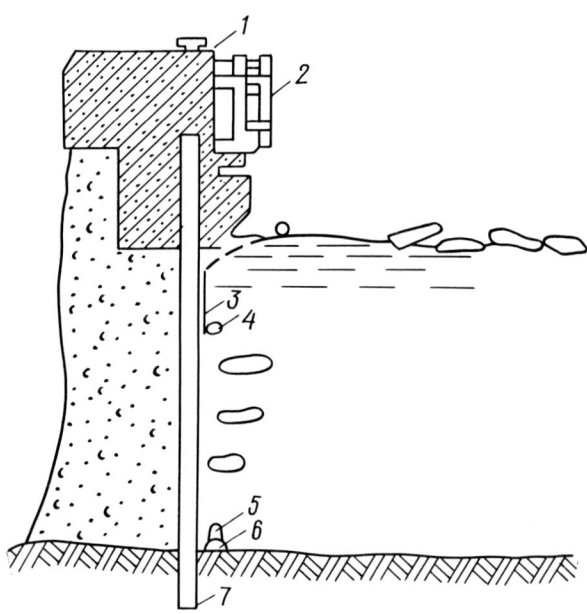

Fig. 2.7. Device for making an ice-free lane along a pier [50].

1 - pier; 2 - deflecting member of pier; 3 - frame with axis of rotation; 4 - pipe with heating medium; 5 - bubbler; 6 - bubbler foundation; 7 - air line.

pipelines for discharging hot gas if there is a total lack of heat in the water [46].

A device (Figure 2.7) comprising a rotating frame hinged to the wall of the dockside, a pipe with a heating medium fixed on the frame below the maximal water level, and a bubbler resting on the bottom, may be used to maintain a part of the water surface free of ice.

When assembling or repairing the upper pipe, the frame is turned on hinges and the bottom end of the frame with the pipeline fixed to it lifts free of the water surface, ensuring free access. The frame is suspended vertically in its working position, fitting the dimensions of the deflecting members.

The device operates as follows. Compressed air is supplied into the bubbler to create a vertical water–air current along the dockside. The vertical current flows over the heating pipe and is warmed either by heat transfer through the pipe walls, if the latter are solid (without performations), or by the discharge of the heating medium through the performations. The vertical preheated current becomes horizontal at the water surface and flows away from the dockside in an integral surface current, performing two tasks simultaneously: preheating of the surface layer in the water body, and removal of broken ice from the wall by a unidirectional strong current. At the same time, icing of the bearding and the formation of an icepack on the dock wall is prevented.

An ice-free lane can be maintained by means of a device consisting mainly of a submersible gas burner. The burner is placed in a cylinder with an open top and apertures for the ingress of cold water. The water is heated in the cylinder on contact with the products of combustion and with the gas; it mixes with them, and moves through a pipe to the surface of the water body.

An installation developed and tested at the Leningrad Institute of Water Transport comprises seven perforated pipes, each of 37.5 mm diameter and 71 m long, laid 8 m apart along the ways of a slip at a distance of 0.5 m from them. The perforated pipes were connected to a common trunk pipe of 100 mm diameter laid on the shore parallel to the water's edge. The pitch of the outlet apertures was 2 m. The diameters of the apertures were chosen to ensure uniform supply of heat along the pipe, varying from 3 mm at the beginning to 5 mm at the end of the pipes [11].

A centrifugal pump supplied warm water into the pipes. Air pipes with a 32 mm diameter and 2 mm outlet apertures, arranged 2 m apart, were laid on the bottom of the water body alongside the water pipes. Compressed air was supplied into the water pipes from an industrial air trunk line to prevent the formation of ice plugs when operation of the installation was stopped.

Testing was carried out in two stages. In spring, the average flow of warm water was 39 m^3/h at the beginning and 22-27 m^3/h at the end of the tests and the maximal flow of compressed air was 14 m^3/min (under atmospheric pressure). Testing was performed under the following conditions: average ice thickness 0.72 m, average water temperature in the water body +0.1°C, average atmospheric temperature -4.8°C.

Ice was melted during the tests on an average area of 1500 m^2. The maximum efficiency of the installation was 39%, the minimum - 21%.

In autumn the tests ensured an ice-free water area at the slip when lifting ships, which proved the effectiveness of creating and maintaining ice-free water areas in a water body lacking natural reserves of heat. However, the installation is bulky, and requires double piping (separate pipe runs for the heating medium and for the compressed air), which complicates assembly and maintenance of the piping.

The Leningrad Marine Research and Design Institute developed in cooperation with the Leningrad Institute of Water Transport a method for creating ice-free water areas which avoids the disadvantages of double-pipe installations [50]. The essence of the method is as follows. A steam—air (or gas—air) mixture prepared by an injector-mixer on the shore is supplied to the bottom of the water body. Steam is the heating medium, and compressed or atmospheric air is the bubbling agent. The use of a steam—air agent is advantageous for several reasons. Mixing of the heating medium and bubbling agent enables a single system of pipes to be used, which simplifies the system, reduces manufacturing and maintenance costs, and also facilitates assembly of the installation. It is unnecessary to use pumps and boilers, because a jet injector-mixer apparatus carries out their functions. An additional advantage is the supply of a uniform mixture of the heating medium and air into the water body via a single-pipe system.

Hydrodynamic plants include devices to create a directional high-speed water flow in the form of a turbulent jet. A jet may be generated by mechanical means (flow generators), pneumatic means (ejector or air-lift), or hydraulic means (water-jet apparatuses).

Flow generators are most commonly used, for example screw pumps creating a high-speed flow. The screw is rotated by an electric motor encased in a waterproof housing. The destructive effect of the flow on the ice cover depends on heat transfer from the water flow to the ice, and the erosive effect of the turbulent jet on the ice. An empirical relationship is used to calculate the size of the lane made by means of flow generators; only the thermal effect on the ice is considered. The length of the lane is determined by the formula

$$L_1 = 127 \times 10^3 R_o U_o t/s,$$

where R_o is the flow generator nozzle radius (m), U_o is the average jet velocity at the nozzle outlet (m/s), t is the average vertical water temperature in a body at the site of the flow generator (°C), and S is the heat transfer from an open water area. The lane's width $b_1 \approx (0.12 \ldots 0.14) L_1$.

The flow generator produces two flows: a jet discharged from the extension piece, and a flow in the direction of the suction hole in the flow generator. Both flows make a circulation current similar to the one formed by the operation of a pneumatic unit. As a result, the return flow brings with it the lower, warmer water mass. When this contacts the ice, the warm water raises its temperature and melts it. The depth of the layer involved in the circulation depends on the intake hole and the capacity of the flow generator. It can be increased by means of special pipes to raise the water from the required layer.

The flow also has a hydraulic erosive effect on sea ice, which contains a large number of cavities filled with brine. During field tests in the McMurdo Sound, a device called the Aquatherm was sunk under a 2.4 m ice sheet to a depth of 3 m. Two electric motors with a power up to 7360 W mixed the water, which had a temperature of -2.2°C, at a rate of 10-15 m/s. An area of about 8×24 m was totally ice-free about eight days later. A lane 14 m in diameter was made in the same region in an other experiment. About 3.18 t of ice was melted in 4.5 days [223].

In the case of a flow generator with a pipe the latter is assembled in several sections so that when the bottom section is lifted by a cable, the others telescope into one another. The top section is connected to the suction cavity of the flow generator. The flow generator is sometimes mounted on the side wall of a floating dock in such a way that it can be displaced on a vertical rod and turned through 180°. The pipe folds when the dock is flooded. This type of flow generator has several functions. It can make lanes upstream from the dock and along its side wall even in a thick ice cover, and protect the dock's face against broken ice. These devices, which are deployed in parts of the water body where the bottom water has a sufficiently high temperature, ensure an ice-free water area. If an above-water flow generator might be damaged, an underwater model is used. The flow generator can be water-cooled by the suction flow, or air-cooled (pat. 27,407 Finland); in this case the air is sucked through a hose, driven by an impeller motor through the body, and discharged via a hose to the atmosphere.

The direction of the flow generator's stream has a considerable influence on the effectiveness of its operation. An upward stream (pat. 3,083,538 USA, 1,166,706 FRG, 1,237,407 France) reaches the water surface but is effective on only a small area of the water body. A pump is placed on the bottom of the water body to draw in more water from the warmer bottom layers. The pump supplies water along a pipe which is laid parallel to the surface with outlet nozzles directed upward. The pump engages and disengages automatically, depending on the surface water temperature; it is controlled by a special device connected to a surface water temperature pickup.

This method of lifting warm water expands the operating area of one flow generator, but it has the following disadvantage: when the depth is great the streams discharged from the nozzles are unable to break through the water thickness because of their low kinetic energy and the small velocity of each stream. This disadvantage can be eliminated to a certain extent by arranging a propeller pump on the bottom of the water body (pat. 3,109,288 USA). Air is supplied to the pump's screw from the atmosphere. The pump produces a mixture of water and air, with a density less than that of the water. The water—air mixture rises to the surface, carrying with it warm bottom water. A screen with a fine mesh at the pump outlet divides the discharged air stream into fine bubbles.

An ejector device may be used to make a small ice-free lane in water bodies with running water and strong surface currents (Figure 2.8). Warm water from the deeper part of the water body is ejected via a pipe because of the passage of a high-speed surface current through a nozzle and a diffuser. The pipe with the ejector has a rudder-float in its upper part, and the bottom end is fastened to an anchor. The device is useful only on rivers with a strong current, or water passageways of upstream hydroelectric power plants. It is unnecessary to supply and use electric power in this case.

Flow generators with flow preheaters (Authors' Certificates 242,691 and 318,510 USSR), have been developed for water bodies without natural reserves of heat. The temperature of the flow is raised by supplying a heating medium to the suction cavity of the flow generator, or by heat transfer from the heating elements in the discharge cavity of the pump. The heating medium (steam or preheated water) is supplied through perforated pipes. A steam—water ejector with a diffuser facing the scrub hub is mounted instead of perforated pipes in the suction cavity to increase the effectiveness of the operation. If a heating medium such as steam or hot water is unavailable, the flow may be preheated by means of electric heaters built in to the plates of the flow-generator's straightening apparatus.

The hydrodynamic method with flow generators is useful in the same water bodies as the pneumatic method, but its use is limited by the greater consumption of power and the smaller width of the resulting lane.

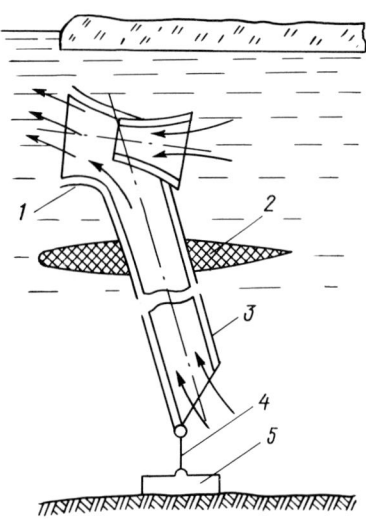

Fig. 2.8. Ejector device for making local lanes [50].

1 - diffuser with nozzle; 2 - buoyancy; 3 - suction pipe; 4 - cable; 5 - anchor.

Flow generators can be used effectively to prevent ice formation, or to break up ice on small areas, as well as to hold up and move away drift ice.

A contact-free device has been developed for floating docks to remove broken ice from under a ship's bottom. This device comprises an underwater flow generator fastened on brackets in the diametral plane of the dock on the working face of the pontoon, capable of turning in the vertical plane. A flexible cable tie, fastened to the face of the pontoon and adjusting the maximal turning angle of the pump, is attached to the face of the flow generator on the side nearest its motor (Author's Certificate 266,608 USSR).

When the dock is flooded, the pump is arranged horizontally and in this position it cleans the water area near the dockface of floating ice. The extreme position of the pump is first adjusted by a cable to ensure an optimal attack angle of the flow in relation to the bottom of the ship passing over it. A directed water flow washes away the floes from under the ship's bottom and drives them away from the dock.

A device with a similar principle of operation has been developed for dry docks, but it has greater effectiveness and range of application (Author's Certificate 312,786 USSR). It comprises an underwater flow generator, which with the aid of a double-arm rod is placed on the dock bottom in its diametral plane near the gate. When the ship approaches the pump the reaction of the flow increases owing to the impact against the bottom plating of the ship; the pump automatically moves away a little, although always remaining near the ship's bottom. The sinking of the pump by the effect of the flow reaction is achieved by a ratchet-and-pawl mechanism, which is controlled from the dockside to ensure surfacing of the pump when the ship passes over it (Figure 2.9). A great

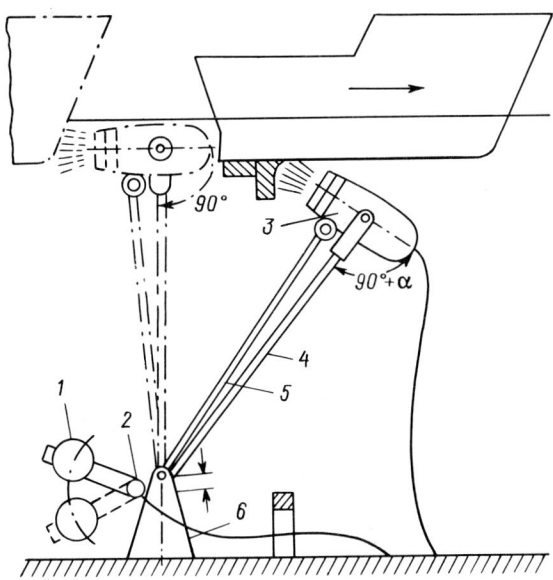

Fig. 2.9. Hydrodynamic device for removing floes from under a ship in a dry dock [50].

1 - counterweight; 2 - ratchet-and-pawl mechanism; 3 - pump; 4 - rocking double-arm rod; 5 - tie; 6 - foundation.

advantage of both the hydromechanical devices described above is total automation of floe removal from the underside of ships of various configurations and elimination of rigid contact between the de-icing devices and the ship's hull. Erosive thermal destruction of the ice cup on the underwater part of the ship's hull by means of water streams is used in ship-repairing. Attempts have been made to use tugboats placed next to the ship's side to break the ice cup by streams from their propellers. However, this process is very lengthy and not highly effective if the ship has bilge keels. Ice is also broken by a stream from an underwater flow generator fastened to a telescopic rod on the ship's stem (Figure 2.10). The flow generator is held on guy cables so that the axis of the flow generated coincides with the baseline of the bottom in the diametral plane of the ship. The flow generator is supplied by the ship or shore power circuit. The erosive effect of the stream in this device may be enhanced by a thermal effect because of the heat of the deep water sucked in by the flow generator.

Preheating of the water passing through the flow generator by steam or electric power has been suggested. A plant has been tested with water passing through the flow generator preheated by hot gas [46]. The plant

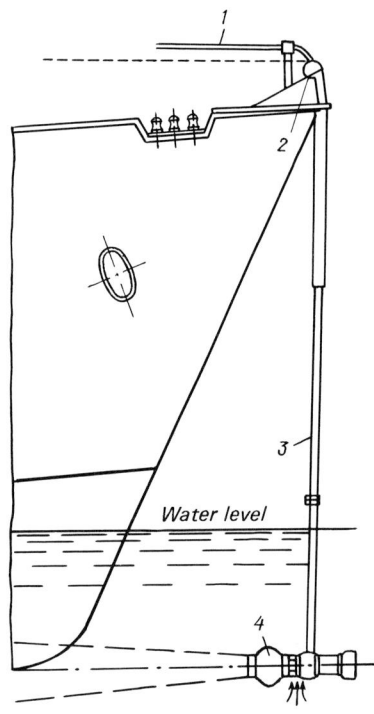

Fig. 2.10. Flow-generating device to remove the ice cup from a ship's bottom [50].

1 - power cable; 2 - base; 3 - telescopic rod; 4 - underwater flow generator. The water level is shown.

comprises a 10 kW flow generator, a compressor with a capacity of 5 m^3/min, and a combustion chamber with an injection nozzle. The tests have confirmed the potential value of developing and using plants with water preheated by hot gas. Comprehensive experience has been gained in Scandinavian countries and Canada in the use of flow generators: in the USSR they have recently been used to maintain lanes upstream of dams and locks.

An essential disadvantage of hydrodynamic methods is the increased heat loss from the water body in winter, which obviously has an adverse effect on its temperature regime, especially in the case of a small or shallow water area.

In Arctic conditions these methods of breaking the ice cover seem rather unpromising, because isothermal processes are observed in Arctic seas (100 m and lower), owing to vertical winter circulation in the upper layers, and the temperature jump layer is at a considerable depth. Experiments in the seas off Greenland and the Canadian Arctic have demonstrated only a slight delay in the formation of an ice cover but there was no further effect [122].

Artificial heating of water areas to melt the ice by discharging steam, hot gas, or specially heated water into the water body, or by burning natural gas, is economically justifiable only to create lanes of limited dimensions [10, 11].

The demand for ice-free water areas is great. One attractive idea is to use industrial warm water, which in most cases is discharged and is of no further use. Some experience has been gained in this respect in France, by organized discharge of waste warm water from power plants into rivers and canals. Thus, parts of the Seine (200 km), Marne (20 km), Moselle (90 km), and Oise (100 km) have been practically protected against freezing [50].

When a 1200 MW steam power plant was planned on the River Elbe (Federal Republic of Germany) the possibility was considered of preventing ice formation by discharging warm water into a 20-25 km downstream stretch of the river. It was calculated that the heat supplied would be sufficient to prevent ice formation under such several (for Germany) climatic conditions as a mean minimal air temperature of -12°C at a wind velocity of 5 m/s. The possibility of heating northern canals has also been studied. It is sufficient to discharge preheated water from a 420 MW steam power plant to keep canals free of ice on 20 km stretches in most severe winters. About 0.5 m^3/s of water at 12 − 16°C is discharged into a canal near Hannover. Observations were carried out in the 1969/1970 winter season of the effect of discharging warm water on the formation of ice in the canal. If warm water was not discharged into the canal it was found that the ice thickness reached 0.3 m. Less ice accumulated between the 6th and the 20th km, in the zone of indirect effect of the warm water. No ice appeared in the zone of direct effect of the warm water, and the ice thickness reached 5-10 cm only under most unfavourable conditions, without interfering with navigation [50].

The slips of ship-repair yards in the USSR have been protected against icing by means of artificial heating on the Neva and Volga rivers, as well as water areas for storing timber at wood-pulp and paper combines [10, 11, 12].

The studies have confirmed the technological potential of using installations for artificial heating of water areas to ensure prolonged operation of the slips in the internavigational period. The results have demonstrated that it is necessary to discharge 50 m^3/h of water at 30°C into the water body to melt an ice cover 0.5 m thick on an 80 x 60 m area in 10 days, and it is necessary to discharge at least 24 m^3/h of water at 30°C to keep the water area of the slip free of ice.

An analysis of these experimental results demonstrates that artificial heating of water areas consumes a great deal of power and is economically advantageous only if there is a source of cheap energy, such as the heat of waste water of industrial enterprises. The melting of the ice is more efficient if the heating medium is discharged near the bottom surface of the ice. This problem is solved quite simply if there is a natural current in the water body which is sufficient to distribute the warm water. If there is no such current it is possible to use a device to form a directional stream (Author's Certificate 394,494 USSR).

Water preheating to maintain ice-free lanes under the severe conditions of the Arctic requires still greater consumption of energy. According to calculations, the heat loss from an open area of an Arctic water body exceeds 41.9 x 12 J/cm^2 in winter even when a loss of heat from the lane is prevented artificially. Hence, it is necessary to burn 24 t of coal of standard caloric power at 100% efficiency of the thermal plant to keep a 10 x 10 m lane free of ice.

It is therefore not feasible to maintain ice-free lanes in winter because of the enormous consumption of energy. Artificial heating in combination with other methods of ice destruction is more economical. It seems advisable to carry out incomplete melting of the ice on some water area, i.e. to supply enough heat to the ice to prevent its thickness exceeding some predetermined value. Subsequent removal of the ice may be accomplished by mechanical or other methods.

New methods for reducing the mechanical strength of the ice cover include an interesting way of artificially creating a layered structure (with air interlayers) in the ice cover, which is easily broken mechanically and is characterized by relatively low heat conductivity [158]. The principle of creating this layered structure is as follows. A lane of the required dimensions and configuration is made in the ice cover; air is supplied under the ice when the first layer has frozen, and then the lane is left to freeze a second layer, and so on. The mechanical strength of the frozen ice cover is reduced mainly because the ice layers, separated by air, break much more easily (Figure 2.11).

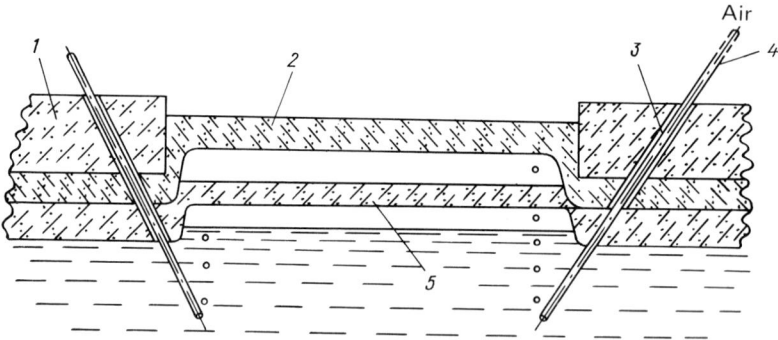

Fig. 2.11. Cross-section of easily destroyed layered ice structure.

1 - ice cover when making a lane; 2 - first layer of fresh-frozen ice; 3, 4 - air ducts; 5 - second layer of frozen ice [158].

The necessity of maintaining the thickness of the layered part of the ice cover the same as that of the undisturbed ice requires less energy consumption. This is because the heat conductivity of the ice is reduced by the presence of air layers which have low heat conductivity. Test results on a freshwater lake near Yakutsk, and on a lake and a seawater area at Vankina Bay (Eastern part of Laptev Sea) have confirmed its potential usefulness.

V.P. Gavrilo et al. (Author's Certificate 958,582 USSR) have suggested a more efficient method of keeping water bodies free of ice, in which increased effectiveness is obtained by means of a multilayer thermal-insulating coating. This is made by setting a layer of ice and sorption material with a low coefficient of heat conductivity mixed with a heat-insulating fluid — oil products for example — and subsequently alternating the layers. When an ice layer freezes on the water surface, a heat-insulating fluid is pumped underneath mixed with the fines of a sorption material. When a second layer has been frozen, the next layer of heat-insulating fluid is pumped underneath, and the operation is repeated as many times as preliminary calculations suggest.

Oil products, which are lighter than water and have a coefficient of heat conductivity 15 - 20 times less than that of ice, can be used as a heat-insulating fluid. Foam plastic, whose coefficient of heat conductivity is 40-50 times less than that of ice, can be used as the sorption material. When there is a temperature difference in the layers of oil products, which are at most 2 cm thick, the transfer of heat through the layers is performed only by molecular heat conductivity (not by convective heat exchange as in the case of air interlayers)).

The effectiveness of a multilayer thermal-insulating coating is characterized by the thickness of the freezing ice, which is calculated by the formulae

$$h_k = -h_t + \sqrt{\left(h_t + h_k\right)^2 + \frac{2\lambda_1(t_k - \theta)}{\rho \cdot L_v} \cdot \tau};$$

$$h_t = \frac{\lambda_1}{\alpha} + \sum_{(i)} h_i \frac{\lambda_1}{\lambda_i}$$

where h_k is the thickness of freezing ice; h_t is the thickness of the thermal-resistant layer; h_k is the layer of earlier frozen ice; λ_1 is the coefficient of ice heat conductivity; t_k and θ are respectively the water temperature at which ice formation occurs, and the ambient temperature; τ is the time of ice formation; L_v is the specific heat of phase transition; α is the coefficient of heat exchange between the surface and the air; h_i is the thickness of the insulating layer; and λ_i is the coefficient of insulation heat conductivity; ρ-ice density.

Theoretical calculations, which did not take into account the heat input from the water, have demonstrated that under Arctic conditions five days are necessary to freeze 16 cm of pure ice, while under the same meteorological conditions 30 days are necessary to make a thermal-insulating coating of 16 layers (the thickness of each ice/straw oil layer is 1 cm). The thickness of the ice on the water body during the entire autumn/winter period will be 2.8 m, while the ice thickness on a water body covered with a thermal-insulating system of 16 ice/straw oil layers will be 1.3 m.

The effectiveness of a thermal-insulating coating comprising ice and straw oil was tested in the Arctic in autumn. The coating comprised three layers of straw oil each about 1 cm thick, and was made in 24 days. The thickness of ice frozen without a thermal-insulating coating was 0.26 m at the number of degrees/subzero days -311.3°C, while the thickness of ice with a thermal-insulating coating was 0.15 m.

The use of heat-insulating fluids with a density greater than that of the ice but less than that of the water provides the coating with mechanical strength and preserves its thermal insulating properties, because in the case of thermal cracking of the ice interlayers the ice fragments will be held by the buoyancy force on the liquid surface, preserving the ice layers against collapse and eliminating the penetration of cold air into the multilayer coating.

The coating is easily made on large water areas at low material cost, because waste combustible and lubricating materials, as well as the waste solid heat-insulating substances, can be used as thermal-insulating materials. These coatings greatly reduce the expenditure of labour in cleaning drainage pits of ice, and also greatly reduce the depth of soil freezing in autumn and winter, thereby considerably extending operational time. This method can be used when extracting useful minerals from the bottom of Arctic seas, to prevent freezing of the operative water areas. When this method is applied, the thickness of the frozen ice under a thermal-insulating coating is 2-5 times less than under natural conditions.

In recent years the idea of using the energy of waves to prevent ice formation in harbours, with the aid of wave generators, has been advanced. The very significant distinguishing property of wave generators is the fact that it is possible to generate waves to prevent ice formation on large areas and distances with low power consumption by very simple means. The wave generators create a mobile wave front, causing continuous displacement of water particles along the vertical line, and thereby preventing the accretion of ice crystals which freeze in the supercooled water. Simultaneously, mixing of the water layers in the depth of the water body occurs, lifting warmer and more saline bottom water to the surface. Also, currents are formed with a low (about 9-10 cm/s) surface horizontal velocity, which cause the forming sludge to mix in the direction of wave propagation. Wave generators are used most effectively in harbours having access to ice-free water bodies, as well as on water areas free of drift ice (Figure 2.12).

Wave generators represent a new type of anti-icing equipment. Only experimental prototypes of low capacity have been tested so far, but the results prove their potential effectiveness. Experiments on wave generators were carried out for the first time in Canada in 1962. The waves were generated by a vibrating ram, which was a wooden beam 10 x 10 cm in cross-section, and 3 m long, with terminal plates and stabilizing weights. It received oscillatory motion from a 0.2 kW electric motor.

The ice cover was 24 cm thick, reducing to 1 cm at a distance of 25 m from the wave generator. After three hours of operation the ice cover was destroyed in a zone 8 m wide and 25 m long, and the water area remained ice-free during the following week despite the low ambient temperature, which went down to -30°C.

The power of driving electric motor in experiments with a buoyant wave generator was 0.35 kW, and oscillatory motion with an amplitude up to 25 cm was employed. Waves of various characteristics could be generated by adding pontoons of different dimensions. At the beginning

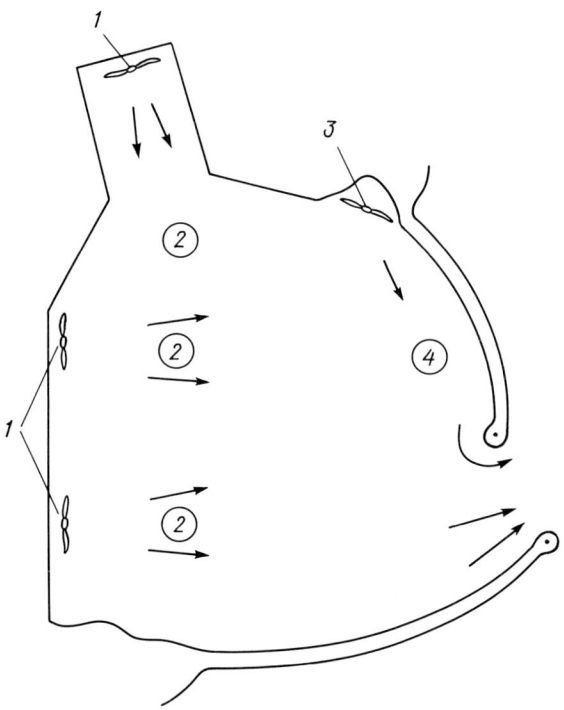

Fig. 2.12. Prevention of ice formation on the water area of a port by means of wave generators [50].

1 - wave generator; 2, 4 - front of propagation of direct and reflected waves; 3 - reflecting screen.

of the operation the thickness of the ice cover was 40 cm at an ambient temperature of 10°C. Operating for four hours, the wave generator cleared a water area with a radius of 100 m. The next day was windy and the entire water surface of the lake was cleared, though the neighbouring lakes were covered with ice for another week.

These results show that a wave generator can be used on water areas with a thick ice cover. Experiments in a shallow trough with cold water close to freezing point have shown that the absence of a temperature gradient with depth reduced the effectiveness of the wave generator only slightly. This leads to the conclusion that wave generators can operate effectively on water areas lacking natural reserves of heat.

Sludge forms on the water surface at low air temperatures, and at a water temperature close to freezing, despite the operation of a wave generator. Experiments demonstrate that the waves displace the forming sludge in the direction of wave propagation, thereby preventing the formation of solid ice. A sludge layer gradually forms on the water surface, greatly reducing the rate of ice formation because of the heat-insulating effect. The sludge has practiclaly no effect on the height of the generated wave.

Various designs of wave generators have been developed for practical applications of this new method of eliminating ice formation. A large-diameter log has been used as a wave generator of very simple design (Figure 2.13). Reciprocating motion in the vertical plane is imparted to the log by a pile-driver.

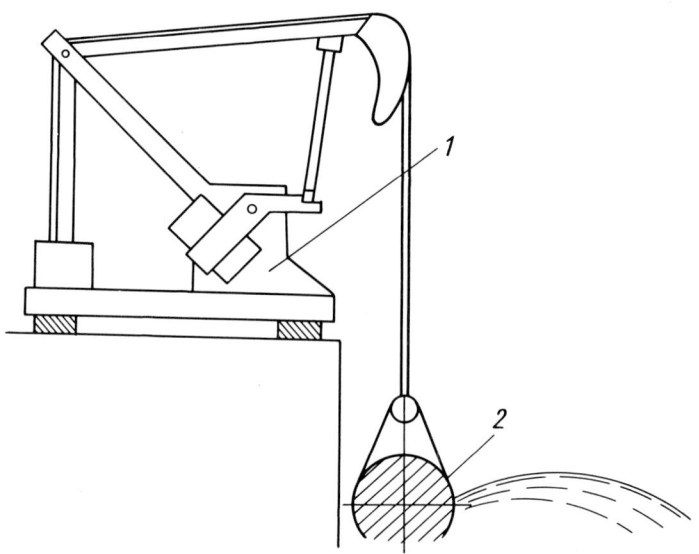

Fig. 2.13. Ram wave generator [50].

1 - driving mechanism; 2 - ram.

Experimental testing has indicated that the operating range of the waves is reduced if the vibrating member is very short, because of the effect of wave diffraction. The diffraction losses are minimized if the length of the wave-generating member exceeds the wavelength by a factor of 5 or 6. It is easier to provide a longer wave-generating member in buoyant units that are mounted on pontoons with extensible supports.

The principle of operation of a buoyant wave generator with a rotary member is given in Figure 2.14. Several designs for wave machine installations are described in pat. 3,477,233 (USA), based on the principle of operation given above.

According to preliminary calculations, a wave generator making waves 1 m high and 10-12 m long removes ice sludge to a distance of several kilometers. The waves create no obstacle to various operations on the water surface. A wave generator with a rotor is most suitable for this purpose. Its power consumption should be about 2.2 kW per metre of rotor length.

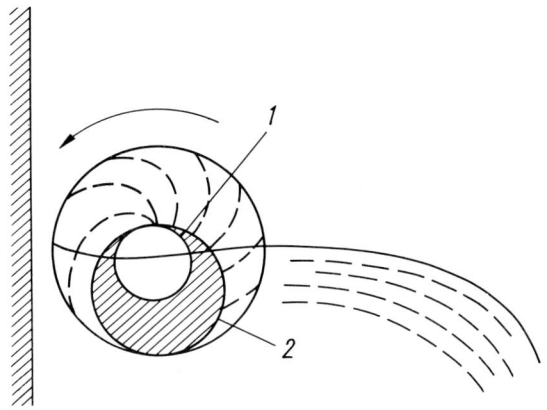

Fig. 2.14. Rotor wave generator [50].

1 - metal pipe; 2 - polyethylene plate.

2.3. Steam—Water—Air Methods

The use of steam, hot water and air for melting lanes in sea ice has been widely applied by I.G. Petrov on the water areas of Arctic ports [129]. A boiler unit was used with a capacity of about 1 t/h at a fuel consumption of about 2.5 t/day. The steam temperature was 120°C. The heat was supplied to the ice along a pipeline made up of 51 mm diameter metal pipes. The steam line was laid in two rows to accelerate ice melting. The melting rate could be accelerated by using a perforated pipeline (with holes 1-2 mm in diameter and 1-1.5 m apart), whence the steam discharged and contacted the surrounding ice or water.

To prevent excessive heat losses it is necessary to ensure that the pipeline, especially if it is a perforated one, is always submerged in the ice or meltwater. A closed system is preferable because when using a perforated pipeline part of the steam leaves the system, and it is necessary to add fresh cold water, which creates heavy operating conditions for the boiler.

Pipelines laid on the ice surface sink after an hour into the ice so that they are almost totally covered with meltwater and most of the heat is consumed through the water to melt the ice. Heat is then lost only from the surface of the forming lane. These losses are quite small and can be reduced in calm conditions if the surface of the snow water forming on the ice is covered. The pipelines sink on the average 20 cm into the ice after 1 h heating, and the meltwater on the ice surface, i.e. above the sea level, drains away, leaving narrow dry grooves over the pipelines and slightly wider than their diameter. According to experimental data, the pipelines sank from 30 to 70 cm after 15 h heating. The strip of meltwater on both sides of the pipelines was 2-3 m wide. The heat transferred from the pipelines through the surrounding water

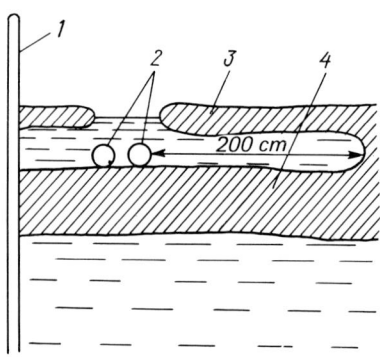

Fig. 2.15. Beginning of formation of snow water on ice near pipelines [129].

was consumed mainly to melt the ice cover. The snow water on ice was formed as shown in Figure 2.15. The pipelines (2) sank through the upper ice layer, then, sinking still lower, started to heat the surrounding water, which had no opportunity to drain away under the ice. As a result, an ice crust about 10-15 cm thick formed over the snow water on the ice. The dynamics of ice melting, depending on time, and the consumption of fuel and steam are given in Table 2.2.

Table 2.2.
Amount of steam consumed for melting a lane [129]

Heating time (h)	Lane		Amount of melted ice (m^3)		Heat required for melting 90 m lane ($\times 10^2$)	Consumption			Coefficient of heat utilization
	Depth (m)	Width (m)	length 1 m	length 90 m		Steam	Heat by steam	Fuel (kg)	
1	0.20	0.05	0.01	0.90	2.722	1	27.3	130	0.10
15	0.50	2.5(2-3)	1.25	112.50	340.2	15	409.5	1950	0.84
24	0.75	-	-	-	-	20	546.0	2600	-
39	1.00	4.0(3-5)	4.0	360	1088	50	1365.0	6500	0.80

Note. The caloric value of the fuel is 20.95 MJ/kg; the mean decade air temperature on melting a lane is -25°C; gentle wind.

The relation of the amount of heat required for melting a lane of the given dimensions to the actual consumption of heat, determiend by the actual amount of steam, consumed for heating the pipelines, is the coefficient of heat utilization.

The coefficient of heat utilization characterizes the degree of utilization of the heat of the unit, allows for the heat losses and indicates the effectiveness of fuel consumption in making a lane. The different values are explained by the changing conditions of pipeline operation. The coefficient is very low during the first hour of melting a polynya, because the pipelines, still uncovered by water or ice, are exposed to low-temperature air, thereby losing a great amount of heat. The coefficient of heat utilization increases greatly after 15 h of operation, because the pipeline has sunk into the meltwater, which is covered by a snow 'cap'. When this 'cap' collapses after 39 h of heating, the heat transfer from the water to the air increases and the coefficient of heat utilization decreases slightly. A graph (Figure 2.16), which can be used to calculate the dimensions, time and speed of making a lane in a practical application of the given method, gives an idea of the correlation of the melting depth and heating time.

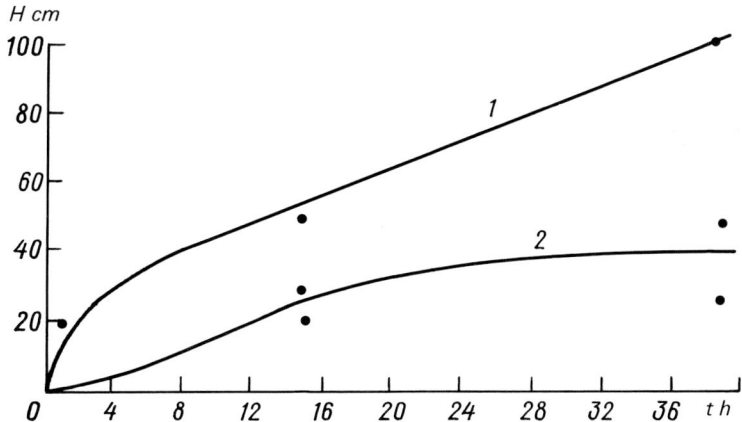

Fig. 2.16. Dependence of depth (1) and width (2) of ice melting by two 90 m long pipelines at steam consumption of 1-1.2 t/h on time [129].

The speed of increasing the lane depth is uneven, and may be divided into two stages: the initial stage (1.5-2.0 h), and subsequent period of heating. The melting rate in the first stage is 10 cm/h; it then decreases and becomes almost constant (2 cm/h).

The width of the lane increases very slowly (0.5 cm/h during the first 2-4 h), and then speeds up to 1-2 cm/h. The first portion of both curves is almost equal, and corresponds to rapid sinking of the pipelines into the ice and slow increase of the lane width. The heat losses are maximal at this time (small coefficient of heat utilization). Only a small amount of ice melts in this ineffective period of unit operation, but then there is an almost uniform and quite marked increase in depth and width of the lane. A lane 2-5 m wide can be made in ice 1 m thick in less than two days.

An experiment steam—air unit has been developed at the Leningrad Marine Research and Design Institute to create an ice-free water area in front of floating docks in Arkhangelsk [50]. The unit (Figure 2.17)

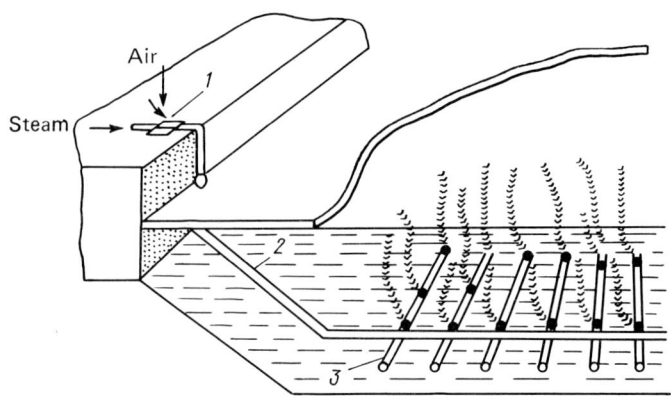

Fig. 2.17. Steam-air unit [50].

1 - steam air injector; 2 - supply pipeline; 3 - distributing pipelines.

comprises three main parts: an injector-mixer, a supply pipeline, and a system of perforated pipes. The main working member of the unit, the system of perforated pipes, comprises a supply pipe, (I.D. 100 mm), and a 70 mm diameter distributing pipeline with eight perforated branch conduits on each side. The inside diameter of the branch conduits is 19 mm, the spacing between them 7 m; there are 13 holes in each branch conduit and the diameter of the holes is 3 mm.

Before the tests the average thickness of the ice at the location of the unit was 0.75 m, reaching 1.5 m in some spots. The thickness of the snow cover was 0.2 m. Steam was supplied with air round the clock at short intervals. The water temperature in the depth of the water body at the location of the unit was as follows: 0.3°C at the bottom, 0.08°C at a depth of 4 m from the surface, 0.03°C at the bottom surface of the ice. The mean air temperature was -8.2°C, the minimal temperature -16.9°C. The amount of heat supplied to the water was 546×10^6 kJ in 327 h of unit operation at an average steam rate of 581.5 kg/h and an excess pressure of 0.4 MPa. High-pressure air was supplied with the steam at a maximal excess pressure of 0.6 MPa. The average air rate (without considering compression under pressure) was 2.98 m^3/min. Ice was melted totally during the test on an area of 986 m^2, and partially melted from the bottom surface of the ice cover to a distance of 35-40 m from the edge of the lane. The total volume of melted ice was 1370 m^3, and the mean efficiency of the unit reached 50.1%.

This method was used by the river fleet to prevent icing of slips [21].

The steam method was also used to remove ice from draining pits in eastern Siberia. The ice was cut into blocks with an area up to 600 m^2 and floated to an excavating machine with a special bucket. A 600 cm dredge with a device for cutting ice by means of steam accelerated the cutting speed by 30-40%, and this reduced the prime cost of the operations almost threefold.

Research and development work has been carried out in the USA to create a new prototype steam-operated ice drill. The construction was based on the design developed by F. Howorka, an Austrian scientist, in 1965 [218].

Propane, developing a higher gas pressure at low temperatures than butane, is used as the source of heat in the drill's burner. The amount of energy produced on burning 1 kg of propane is 5×10^7 J, which is sufficient to boil 18.4 kg of water and raise the steam temperature to 140°C. The steam produced can melt 150.6 kg of ice, or melt a 2.5 cm diameter hole to a depth of 324 m. It is impractical to drill deeper than 16 m, because the heat losses in the hose are too great. The steam is supplied to the ice by way of a flexible hose with double walls (a rubber external wall and a Teflon internal one) through an extension piece with a nozzle (Figure 2.18). A hole 8 m deep and 2.5 cm wide can be drilled in 15 min. The drilling speed drops to 19 m/h at a depth of 16 m, and to 6.6 m/h at a depth of 32 m [216].

A similar drill was used by the 14th Japanese Antarctic expedition in 1972-1974 to melt holes 3 cm in diameter and 10 m deep in firn ice at a drilling rate of about 28 m/h [239].

An apparatus for ice destruction by means of hot-water jets (Author's Certificate 844,465 USSR) has a body of which the lower part takes the form of a set of longitudinal parallel conduits. The upper ends of the conduits communicate with one another, and the bottom ends

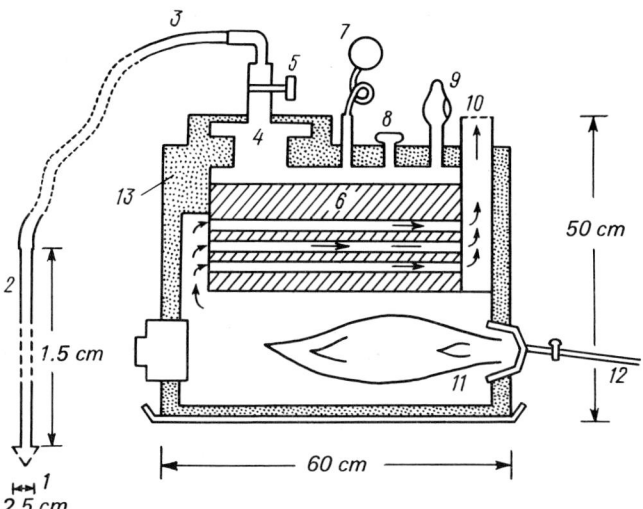

Fig. 2.18. Sectional view of steam-operated ice drull [238].

1 - nozzle; 2 - guide pipe; 3 - rubber hose; 4 - steam; 5 - steam line valve; 6 - water; 7 - pressure gauge; 8 - water filler; 9 - safety valve; 10 - exhaust pipe; 11 - burner; 12 - propane tank; 13 - insulation.

are fitted at the faces with slots and extend outside the body. This design ensures uniform discharge of hot water from the outlet holes irrespective of the way in which the working face of the apparatus is arranged on the ice, and of the ice quality (content of pollutants, presence of cracks, etc.). The latter design makes it possible to develop a high rate of ice destruction (by drilling) at a high efficiency: tests have demonstrated a drilling speed of 40 m/h and the efficiency of the apparatus reached 50-60% with repeated use of the waste water. The testing of experimental prototypes of the apparatus to cut ice blocks out of the ice cover on the Lake Ladoga, and to make trenches in the glacier cover in Antarctica has demonstrated their high reliability and effectiveness.

Steam, hot water, and air from marine engines are successfully used to prevent and eliminate icing of the above-water parts of sea-going ships (Authors' Certificates 280,250 and 300,378 USSR). To this end, the ships are equipped with a stationary system of pipelines, comprising a trunk pipeline with pipes branching to the spots of intensive icing. The pipes have cuts and are fitted with sprayers. The heating medium, which is sea water, discharges through the cuts and sprayers. Sea water is supplied along two lines, one of which is connected to the cooling system for preheating sea water. The latter system comprises an individual heat exchanger, utilizing the heat of the propelling plant's exhaust gases. The steam is supplied to the surface of the ship's above-water part through apertures (Authors' Certificates 195,910; 360,268; 209,224 USSR) and perforated steam lines (pat. 1,315,834 GB).

The formation of an ice cup on the underwater part of a ship's hull occurs due to heat transfer through the plating. The thickness of the ice cup is proportional to the heat transfer rate. Hence, the air temperature inside the ship has the principal effect on the growth rate and size of the ice cup.

It is advisable to close the cargo hatches completely and shut off all the fans in the holds and other openings while awaiting docking at low air temperatures. The ship should be held on an even keel at maximal draught, maintaining a positive temperature in the ballast tanks.

A radical means of preventing icing of the ship's bottom is to equip the ballast tanks with a temporary heating system. To this end, steam is supplied to the water tanks from the shore, a tugboat, or another ship with a steam-boiler plant. The steam consumption is very low, and one ship supplying heat can meet the heating requirements of a considerable number of other ships lying at anchor.

A system patented in Great Britain for heating the hulls of ships lying at anchor are of some interest (pat. 894,878 GB). The system comprises screening devices extending down into the water from the sides of the ship. Perforated pipes for supplying hot water under pressure are inserted between the screening devices under the ship's bottom. To reduce heat losses, the screening device is made of two superimposed tarpaulin canvases with an air space in between. The device is very complicated in manufacture and operation, so it cannot be recommended for preventing the formation of an ice cup, but the ideas it demonstrates exemplify the current state of the art and can be used in developing new techniques.

THERMAL DESTRUCTION OF ICE

The problem of devising active methods for eliminating the ice cup remains current because it is not always possible to prevent icing of the underwater part of the ship's hull.

A widespread method of eliminating the ice cup is to supply steam along a special system of steam lines into tanks partially filled with water. Depending on the thickness of the ice cup, the period of supply of steam is variable and may be five days or more. This is because the ice melts a little, and is then pressed to the ship's bottom again by hydrostatic pressure, while the heat transfer from the hull to the ice through the interlayer of water deteriorates.

Heavy-duty compressor and heating units are necessary to heat the plating of a hull with hot air. The major disadvantages of this method are the loss of much of the heat in the air, which is the heating medium, owing to its discharge from the tanks to the atmosphere, and unproductive use of heat for heating the upper structures of the tanks.

The removal of ice by means of hot water, supplied via a system of perforated pipes assembled on a special stationary buoyant device, necessitates towing the ship to this device and repeatedly moving the ship over the device. The hot water discharging from the perforations of the pipes mixes with the surrounding water, cools, rapidly attains the temperature of maximal density (4°C), and sinks to the bottom of the water body, causing large heat losses [10].

One method for removing ice from the underwater part of a ship involves the supply of steam through a hollow rod with nozzles (Author's Certificate 195,910 USSR). This consumes a large amount of steam, but supplying steam to all points under the ship's bottom is difficult. The method can therefore be used only when there is a large amount of steam at the yards, to remove ice cups from ships of small width and draught.

An effective means for removing the ice cup is a unit designed to maintain ice-free parts of water areas at ship yards: a system using the heat of deep waters, also double-pipe water—air systems, and single-pipe systems supplying steam—air mixtures, which are used for artificial heating of water areas. The ship is arranged over one of these systems. The upward above-zero water currents contribute to effective melting of the ice cup. If the system is large enough, the ice melts simultaneously off the entire surface of the ship's bottom. The ship can go immediately to the dock or the slip, because ship-repair facilities maintain usually ice-free areas at the ship lift or the slip.

Hot-air anti-icing systems are widespread in aviation to eliminate the icing of aircraft, wings, control surfaces, engines, and air intakes. The principle of operation is based on using air which is heated in heat exchangers by the circulation of hot exhaust gases [156].

2.4. Gas-Thermal Methods

Gas-thermal means for ice destruciton are based on using a high-temperature gas jet discharged at supersonic speed from the nozzle of a reactive burner. The burner is a chamber in which the fuel burns and high-temperature and high-pressure gas is generated. The gas has a large reserve of thermal and mechanical energy. The working pressure in the combustion chambers of thermal cutters varies from 0.5 MPa to 0.9 MPa, and the temperature in the chamber reaches 3000-3300°C. The incandescent gas is discharged from the chamber via a nozzle of special shape (Laval nozzle); passing through the critical section of the nozzle, the gas develops supersonic speed [188].

The pressure of the working components is as follows: oxygen 1.2-1.6 MPa, kerosene 1.4-1.6 MPa, and water 0.5-0.8 MPa. The consumption of oxygen for normal burners is usually 8-10 kg/h, and cooling water 200-250 l/h. The efficiency of the reactive burners reaches 60-70%.

Only a supersonic jet ensures effective ice destruction, because high temperature alone is not enough; a large amount of heat must be supplied to the object being destroyed in the shortest possible time. The incandescent jet discharging at supersonic speed contacts the ice surface and conveys a large amount of energy to it in a short time (a fraction of a second). The aerodynamic shocks of the reactive jet contribute to effective and directional destruction of the ice (Figure 2.19). According to experimental data, ice destruction by the gas-thermodynamic method occurs in the form of rapid melting in the operating zone of the high-temperature flame of the burner. Ice destruction in the form of chipping and flaking with removal of the products of destruction, as in the case of destroying hard rock, is not observed. In its appearance, ice destruction resembles the destruction of such materials as concrete and reinforced concrete.

The process of ice cutting is accompanied by a lot of noise, arising from the discharge of the jet from the reactive burner and its impact on the ice surface. During operation a gas cloud forms with a strong odour of kerosene. The meltwater drains and splashes in all directions. From the appearance of the draining water it is evident that some of it is boiling, and the rest is transformed into steam and participates in the formation of the gas cloud. The avergae rate of ice melting by means of oxy-kerosene thermal cutters is 74 cm^3/s.

The capacity of air-reactive thermal cutters is slightly lower than that of oxy-reactive burners, but they feature certain advantages, the basic one being that it is possible not to use gaseous oxygen and water (which is necessary for cooling the thermal tools). This greatly simplifies the apparatus and makes its operation under real working conditions more reliable.

The unit comprises a compressor receiver for feeding air, a fuel tank and a thermal baffle. The temperature of the gas jet at the outlet of the thermal baffle is 1000-1200°C, the jet velocity reaching 1200-1500 m/s at an air intake of about 2.5-3 m^3/min.

Oxy-kerosene and air-petrol combustion chambers differ very little in their effectiveness. Better results are gained with oxy-reactive burners. The reduced cutting area is 7.8 cm^2/s for oxy-reactive thermal burners, and 6.2 cm^2/s for air-reactive ones. The burners of the first type demonstrated a slightly faster ice melting rate, 74 cm^3/s, while in the case of air-reactive thermal cutters it was 52 cm^3/s. However, despite the slight advantages, cutting ice with oxy-reactive combustion chambers has some essential disadvantages. The main one is the necessity

THERMAL DESTRUCTION OF ICE

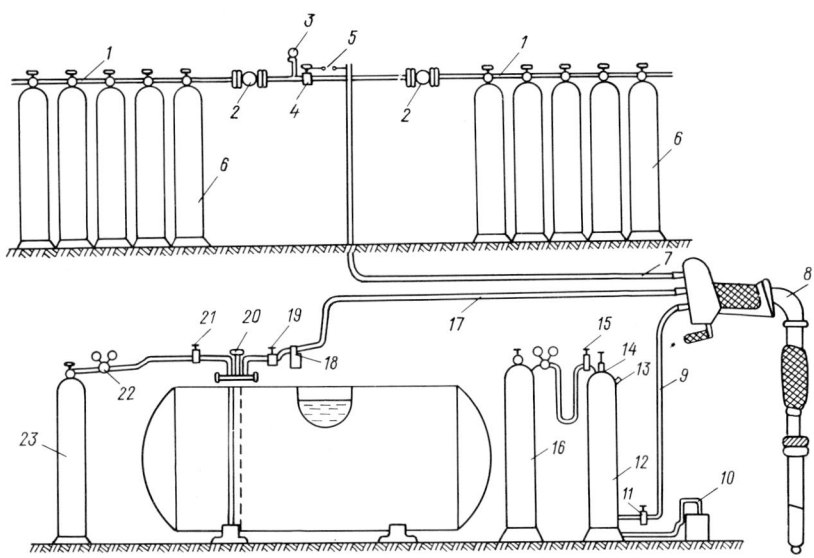

Fig. 2.19. Gas thermal unit with oxy-kerosene burner and water cooling [188].

1 - header; 2, 4 - shutoff valves of header and oxygen take-off;
3 - pressure gauge; 5 - manifold pressure regulator; 6 - oxygen cylinders;
7, 9 - oxygen and kerosene hoses; 8 - thermal cutter; 10 - capacity for drained kerosene; 11, 18 - kerosene and water sumps; 12 - kerosene cylinder; 13 - plug; 14, 20 - check valves ($P = 10.1 \times 10^5$ Pa;
$P_p = 10.1 \times 10^5$ Pa); 15, 19 - shutoff valves; 22 - pressure regulator.

of using water for cooling the combustion chambers in thermal tools, which makes it difficult to operate them in winter. Also, it is economically disadvantageous to use expensive and scarce gaseous oxygen as an oxidizer.

Ice cutting with air-reactive thermal tools is more practical, despite the fact that the ice cutting rate is slightly lower than that for oxy-reactive combustion chambers. Air-reactive thermal tools operate free of failure at low air temperatures, owing to the use of air for cooling instead of water. The use of air greatly simplifies the design of the combustion chamber, making its operation more reliable and simple. Finally, the use of compressed air instead of gaseous oxygen as the oxidizer makes the process much more economical.

When determining the range of practical problems of ice destruction that may be solved by the use of a reactive gas jet, it must be borne in mind that ice destruction by melting is most disadvantageous from the energy point of view, because ice is characterized by a high specific heat of melting (80×4.19 J/kg) as compared to other substances.

Therefore, the range of tasks for which the method is suitable is reduced to individual problems where the volume of work is small and speed is essential.

A hot reactive-gas jet is used, for example, to clean airport runways of snow and ice [91, 140], and to drill holes in the thickness of an Antarctic glacier [126]. In cleaning a runway of snow, gas at 600°C created a heat up to 29.43×10^3 H and left a clean strip 50 m wide behind the machine at a rate of 3 - 10 km/h. The rate for cleaning a runway of ice slush is much less. Data have been published on the use of TG-16 gas-turbine engines and AN-26 aeroplane engines (that have reached the end of their service life) for cleaning roadway coverings of ice [91]. The ice was first melted a little and then removed to a distance by a gas—air jet. It has been calculated [73] that it is necessary to use 1200 l of petrol to remove 15 mm of sleet from a strip 3.66 m wide and 1 km long at the maximal efficiency (40%) of a mobile self-contained melting machine.

American workers used the reactive-thermal method in 1977 to drill the Ross Ice Shelf in Antarctica. They drilled a hole with a diameter of 30 cm and 427 m deep in 7 hours.

A gas-thermal burner for flame drilling looks like a jet engine with an internal combustion chamber encased in an external pipe (Figure 2.20). Cold water circulates between the chamber and the pipe and atomizes around the burning gas jet, discharging under high pressure at supersonic speed from a converging nozzle. The drilling system comprises two air compressors, each driven by an individual diesel engine. The first 208 kW compressor develops a pressure of 0.98 MPa, and the second one boosts this to 8.4 MPa. The braking system of the drill comprises a hydraulic hoist and a mechanism to control the cable. The second (boosting) diesel engine starts the hydraulic pump, the high-pressure water and fuel pumps. The drilling speed in the upper layers of the ice was not more than 0.61 m/min and reached 1.83 m/min only in the last few hundred meters. A heating cable with a power of 60 kW was immediately sunk into the drilled hole to the entire depth of the water column, thereby delaying its closing up for four days [199].

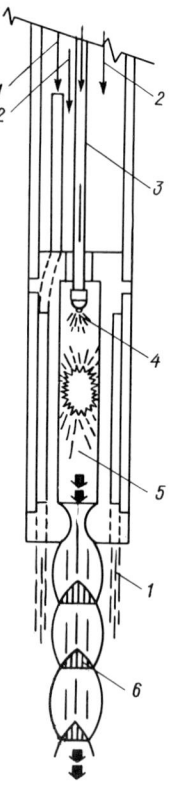

Fig. 2.20. Gas—thermal burner for flame drilling [199].

1 - water; 2 - air; 3 - oil fuel; 4 - atomizer; 5 - combustion chamber; 6 - ignition.

THERMAL DESTRUCTION OF ICE

The gas-thermal method of ice destruction has been found potentially useful, and it has been improved at Memorial University, Newfoundland, Canada, and the Scientific Corporation of Oceanic Research.

The use of ammonia, hydrogen chloride, sulphur chloride, and gaseous ammonium chloride has been suggested: gaseous ammonia is considered to be most suitable. In the presence of water, ammonia first forms NH_3H_2O, an unstable hydrate, which partially decomposes back into gaseous ammonia and water. A powerful heat source is necessary to maintain the compounds in a gaseous condition. The system tested had a heat source of 3 MW providing a gas pressure of 20 kPa [240].

The system comprised cylinders with ammonia and compressed air, pressure regulators, and a long mixing pipe whose nozzles discharged a

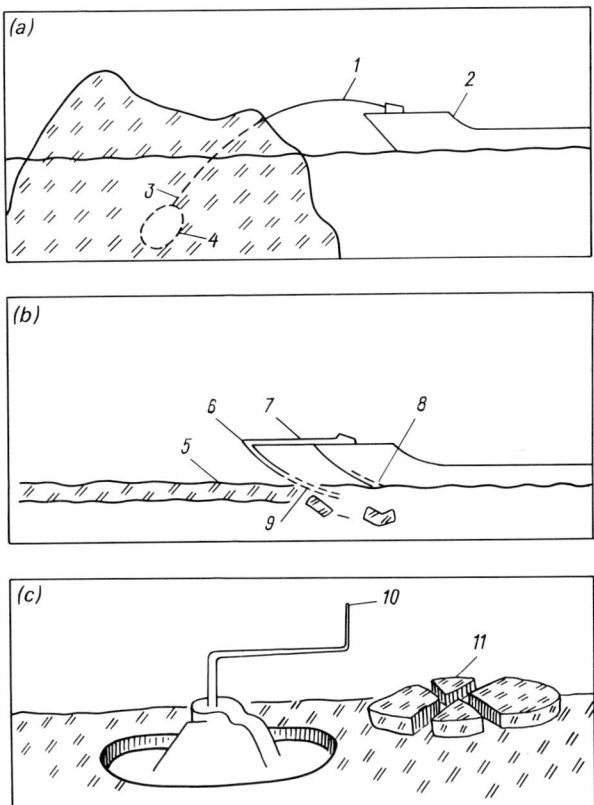

Fig. 2.21. Gas—thermal destruction of ice (a) to make a cavity in an iceberg by blasting and towing, (b) to facilitate icebreaker navigation, and (c) emersion of a submarine [210].

1 - gas-thermal drill; 2 - ship; 3 - drilling trajectory; 4 - cavity in iceberg; 5 - ice; 6 - mixing pipe; 7 - brackets; 8 - perforations in pipe; 9 - outlet nozzles; 10 - rotating gas—thermal bit; 11 - ice plug.

mixture of ammonia and air. The melting rate reached 210 cm/min. The 'saline' effect, which was observed when gaseous ammonia dissolved in the melt water, protected it from freezing.

Another device for flame drilling (pat. 1,315,921 GB) comprises a ram for generating hot gas, and an aperture for directing the hot gas around the circumference of the ram.

A device for drilling ice with a hot gas generator (pat. 2,166,393 France) comprises a combustion chamber and outlet nozzles that ensure heating of the drilling zone around the bit.

The use of the gas-thermal method has been suggested for drilling blast-holes in icebergs to seat charges with the aim of subsequent blasting, or for fixing towing lines, removing ice from the surfaces of ship's structurs, or weakening an ice cover by an outburst of gaseous ammonia under the stem and bottom of a vessel [210]. A rotating mixing pipe can cut holes in ice by means of a gas-thermal jet to allow the surfacing of a submarine (Figure 2.21).

2.5. Electrothermal Methods

Various electrothermal techniques have been developed for the destruction of ice. Among them are the electric thermodrills that have recently become widespread. They are used for vertical drilling of glaciers 1.5 km thick. The method of electrothermal cutting of ice covers has previously been used to free ships from ice before resuming navigation after lying at anchor in winter. Some anti-icing devices are based on the same principle, i.e. local melting of ice by means of an electric heating element. The latter is either a constantan wire supplying electric current, or a working member with high heat conductivity housing a conductor heated by electric power, etc.

<u>Electrothermal drilling</u>. The technology of high-speed drilling of holes in ice by thermal drilling is being intensively developed. The reason for the interest in this method is the high cost, weight, and difficulty of delivering and installing equipment for mechanical rotary drilling.

The essential element of the electric thermodrill is the heater, which melts the ice. Conical and face-type heaters, designed for making holes with coring, were tested during the 6th and 8th Soviet Antarctic expeditions [93]. A circular heater from the Moscow Mining Institute (ETBLK) was tested during the 11th Soviet Antarctic expedition, and a thermodrill from the Leningrad Mining Institute (TELGA-2), tested during the 13th Soviet Antarctic expedition, was used to drill a hole 212.5 m deep [94-96]. The shelf ice in the area of the Novolazarevskaya station was drilled through for the first time by the 20th Soviet Antarctic expedition. The hole depth was 347.0 m, the diameter 9 cm, the core diameter 8 cm, and the drilling rate 5 m/h [71].

The use of an aqueous—alcoholic solution for hole plugging made it possible to simplify greatly the temperature changes on cold glaciers. The concentration, temperature and density of the solution was in accordance with the ice temperature. The density of the solution increased with increasing hole depth in the presence of a positive temperature gradient. Owing to the total absence of convection, the solution in the hole preserved a stable equilibrium condition and its temperature was equal to the ice temperature in all the parts of the hole. Tests of various fluids for plugging holes, carried out at North Pole-19 station and on the Severnaya Zemlya Island, demonstrated that the aqueous—alcoholic solution remained stable for a long time; the temperature measured in a plugged hole and a dry hole, was practically the same. Control holes plugged with diesel fuel demonstrated temperature deviations caused by convection of the fluid.

A single-conductor armoured cable (KEP-1) and a wire based on this cable, with the armour replaced by a copper screen, have been developed for thermodrilling. The thermodrills are supplied with single-phase ac or dc, 220-380 V.

Electric needles and core thermodrills (Figures 2.22 and 2.23) are simple and reliable in operation, but thermodrilling requires some skill and attention, especially when drilling holes in snow-firn areas and in highly polluted ice, when it is necessary to reduce the power to prevent burning of the crown bit owing to worsening heat-transfer conditions in the porous parts of the snow and firn, or when drilling into highly polluted ice. Owing to pollution, the drilling rate is 20% less than the design figure when drilling holes under real conditions [95]. TELGA-14MP is a tool which has been developed specifically for drilling-melting ice under field conditions (Figure 2.24). It differs from the TELGA-14M tool in the length of the core barrel.

Drilling-melting occurs under the effect of heat emitted by the crown bit-heater. The melt water is sucked away by the turbocompressor through water-lifting pipes to the water receiver. To maintain a constant pressure on the tool, the brakes of the hoisting winch are adjusted so that the cable runs off the winch drum for 10-15 cm above a critical tension. The frequency of cable run-off indicates the drilling rate and the state of serviceability of all systems of the tool. This has been confirmed by the practice of drilling ice deposits by electric thermo-drilling tools on a cable [64, 76].

All the mechanisms and the electric thermodrilling tool are supplied from a 16 kW power plant. Running-in and pulling out of the tool is performed by a 3 kW winch. All the tool systems are supplied via the cable, which is also used for round-trip operations. All the equipment is assembled on a mobile drilling rig, specially designed for drilling holes during scientific expeditions.

Table 2.3 presents some data on the performance of electric thermodrilling.

The design of an icebreaking drilling ship (pat. 3,759,046 USA) has been developed in the USA for operation in Arctic waters. The ship is equipped with a patented trim and heeling system to keep station over the drilling location with a maximal 5-7% deviation from the depth of the location in the presence of an ice field 1.8 m thick, encroaching at a speed of 2 knots.

US companies have developed a design for a towed drilling rig on an air cushion. The unit comprises a shallow-draft pontoon with the drilling equipment and an external heating system, which melts ice 1.5 m thick by means of the exhaust gas from three diesel engines.

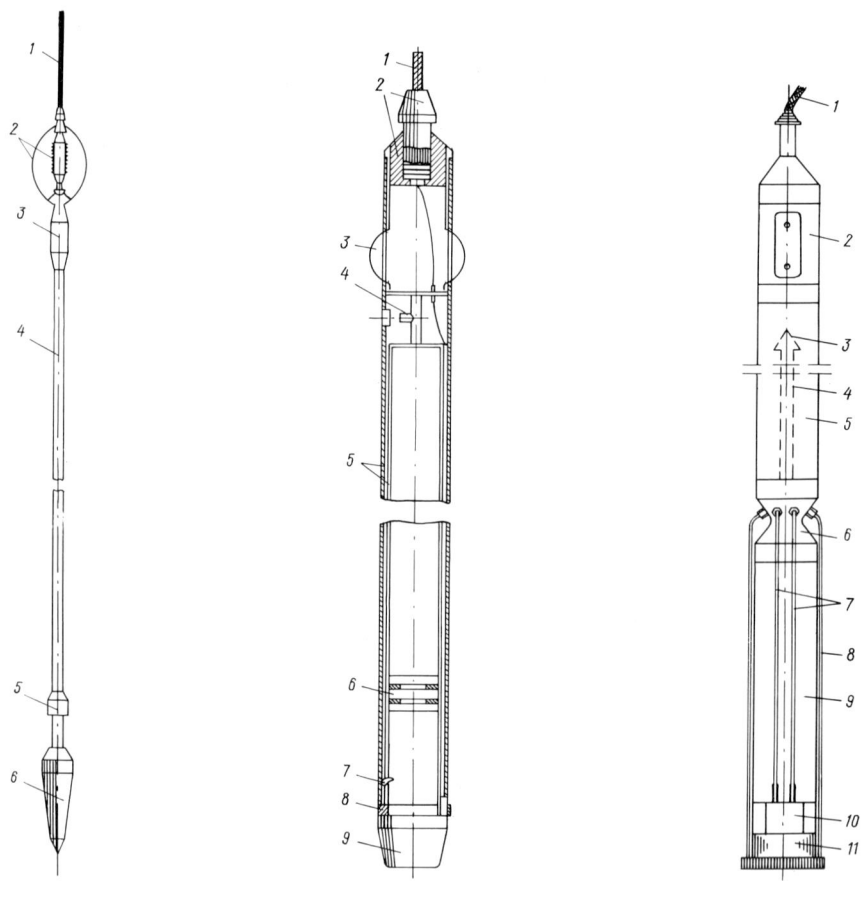

Fig. 2.22.   Fig. 2.23.   Fig. 2.24.

Fig. 2.22. Electric needle (thermodrill ETI-1 for through drilling) [95].

1 - cable; 2 - centring device; 3 - cable lock; 4 - body; 5 - connector; 6 - conical heater.

Fig. 2.23. Electric core thermodrill ETB-3 [95].

1 - cable; 2 - cable lock; 3 - centring device; 4 - charging point; 5 - double core barrel; 6 - piston; 7 - core extractors; 8 - bit flange; 9 - electric heating crown bit.

Fig. 2.24. Electric thermodrilling tool TELGA-14 MP [64].

1 - cable; 2 - cable compartment with turbocompressor; 3 - water deflecting baffle; 4 - central water-lifting pipe; 5 - water receiver; 6 - adapter; 7 - current-carrying wires; 8 - water-lifting pipes; 9 - **core rece**iver; 10 - core extractor; 11 - crown bit-heater.

Table 2.3.
Performance of electric thermodrilling of ice

Thermodrill	Drilling depth (m)	Hole diameter (cm)	Core diameter (m)	Drilling rate (m/h)	Heater power (kW)	Reference
Thermodrill ETBLK		15	9	2.5	5.5	[96]
Thermodrill TELGA-2	190	15.4	7.1	1.6	3.0	[96]
Thermodrill TELGA-3	217	18.0	11.4	1.5	6.9	[96]
Thermodrill ETI-1 for through drilling (electric needle), (Arctic and Antarctic Research Institute	310	4.5	-	5.0-15.0	1-3	[95]
Electric core thermodrill ETB-3 of (Arctic and Antarctic Research Institute	447	12.0	-	2.0-6.0	1-3	[95]
French electric thermodrill	-	14.0	10.2	3.6	6.0	[213]
Electric thermodrill (University of Minnesota)	-	-	-	7.6-8.2	-	[217]
Electric thermodrill (Washington University)	210	-	15.0	6-8	2.2	[248]
Thermodrill (CRREL)		16.4	12.2	1.9	1.9	[242]

Equipment has been developed in France for thermodrilling large holes in massive ice (pat. 2,338,125 France). A truncated cone with a pump to remove the meltwater is enclosed in a metal cylinder. The electric heating element is arranged at the joints of the casing. Glass-fibre casing pipes, impregnated with epoxy resin, are used to drill an ice cover more than 1000 m thick at an ambient temperature of -50°C [3].

Electrothermal cutting. This is performed by means of a heating wire, supplying dc or ac from a suitable power source. The common working voltage for cutting ranges from 50 to 120 V.

A transformer with a separate reactive coil of the ST-2 system and designed for 250 A maximum, is most reliable and suitable for electric cutting. The mass of this transformer is 100 kg, and the power 15 kW at 180 A. The mass of the voltage regulator is 80 kg.

Ice-cutting apparatus is usually made so that it is possible to melt one or two through cuts in the ice cover at a small angle sloping downward (Figure 2.25). When two slots are cut, the forming cantilever breaks off and can be slid under the ice cover [69, 70].

The cutting apparatus is moved during operation every 15-30 s with 10-15 s intervals. When it is immoble, the weights tension the heating

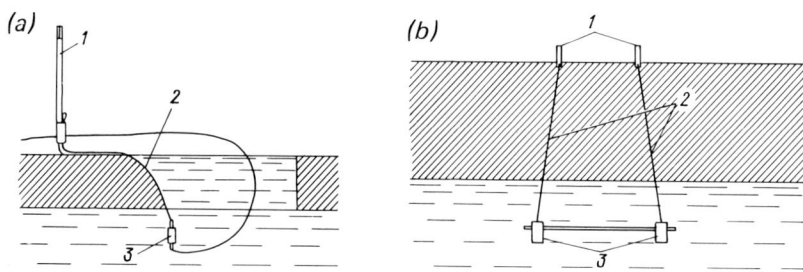

Fig. 2.25. (a) Single-row and (b) double-row electrothermal cutting of ice [70].

1 - holders; 2 - heating wires; 3 - weight.

wires which melt the ice, going down under the effect of the weight. The cutting wire should be sufficiently elastic and possess mechanical strength at the operating temperatures. Tests demonstrate that braids of the required cross-section made of Ni—Cr alloy wires of about 0.5 mm diameter are the best. It is desirable that the length of the heating wire should exceed the ice thickness by 30-40%. In the majority of cases the heating of the part of the wire which is in the water is insufficient, while the wire in the air superheats and burns. Hence, during cutting the entire heating wire should be immersed in the water, or else a composite wire should be used with a much greater diameter in the upper part which is exposed to the air.

A hole is made by means of an electrothermal ice melter (Figure 2.26). The heating wire is placed between two copper sheets, which have

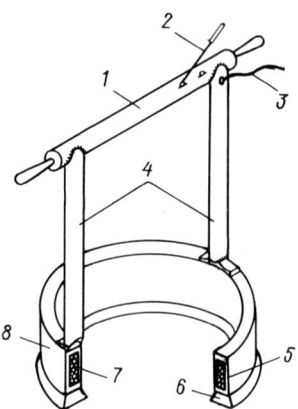

Fig. 2.26. Electrothermal copper melting [148].

1 - handle; 2 - knife switch; 3 - insulated wire; 4 - posts; 5 - electric insulation; 6 - cutter; 7 - heating wire; 8 - copper sheet.

a base in the form of a cutter in the bottom part. Electrical insulating material, for example mica or asbestos, is inserted between the heating wire and the copper sheets. The heating wire is connected to a power source.

The electric power consumed to form a hole in the time t is calculated by the formula

$$N = km(cT + L)t^{-1}$$

where m is the mass of ice being melted, c is the specific heat of ice; L is the specific melting heat, T is the mean ice temperature and k is a coefficient relevant to heat loss (according to data of N.M. Konovalov, k = 1.1...1.2).

For example, a power source of N = 35 kW is necessary to make a hole 350 mm in diameter in an ice cover at a cutting width of 10 mm (air temperature T = -15°C) in one minute. The same hole is made in half a minute by an ice drilling unit at N = 6 kW. In the electrothermal method of ice-cutting on rivers, the speed of cutting by an incandescent conductor, manufactured on nickeline with a high specific resistance, reached 5-32 m/h at a cutting width up to 5 mm and an ice thickness of 0.55 - 1.2 m, depending on the power consumed.

An experimental prototype of an electric thermomechanical hand tool, operating in the high-frequency range, has been tested successfully when drilling slant holes in artificially frozen floating soil in the construction of the Metro [90].

Several recent accounts have dealt with determination of the ice melting rate when applying the electrothermal method. One group established that the rate of melting through an ice monolith by a heating wire at low pressure was two orders of magnitude lower than the theory would suggest. When the pressure exceeded 10^5 Pa, the cutting rate jumped to the design value [242]. A method for rapid calculation of the ice melting rate under a flat heated disc is advanced in another publication [226].

Electrothermal methods of prevention of icing can be divided into two basic types, one of which is designed to reduce adhesion (melting of the bottom thin ice layer), while the second is intended for total elimination of ice (continuous heating method).

The first method envisages the removal of ice by some auxiliary means. The heater in this case should have a power of 1-2 kW/m². The power of the heater in the second case should be at least 1200 W/m² [1].

The use of electric de-icers on ships is limited by their comparatively high power consumption. For example, it is necessary to spend 150 W of electric power to protect 0.1 m² of surface area. The protection of a trawler would require more than 300 kW. It takes 8-20 minutes to eliminate ice 12-13 mm thick by means of an electric de-icer. An electric de-icing system can be divided into sections, energized alternately, to prevent the build-up of ice. The power required is reduced in this case to 20-40 kW [1]. Electrothermal anti-icing systems are now used in aviation for the protection of engines and air intakes, cockpit glass, propellers and stabilizers of turbine-engined aeroplanes, and to remove slush from runways [156].

Such systems are used at sea for protection of ship superstructures, and hydrotechnical structures on the ice shelf. Electrothermal anti-icing systems are classified as either continuously operating or cyclic. The former protect structures during the entire icing period. If a thin ice layer builds up, a cyclic anti-icing system is energized to melt the ice in the contact layer; the ice then slides off by gravity, or is removed by an air flow, or by spalling [156].

Experience shows that wire heaters (pat. 3,800,121 USA, 1,351,158 GB, 2,202,810 and 2,212,261 France) are easily damaged. It is preferable to use current-carrying coatings, which can be in the form of metal films (metal foil, sprayed metal powder, etc.), or polymers filled with a current-carrying material (metal powder, soot, graphite). These are comparatively cheap and more economical [119].

A ship system, 'Spraymat', has been tested more than once in Norway and England. It is a modification of a British aircraft anti-icing system. The tests ended inconclusively: comprehensive testing was too expensive for private shipowners, and the system was soon forgotten. The essence of the system is as follows. The surface area to be protected is coated with a layer of sprayed plastic material. An aluminum alloy is sprayed in 7.5 cm wide strips, and then a layer of pigment anticorrosive plastic is sprayed. The ends of the aluminum alloy strips extend beyond the surface and are connected to a power source, thereby creating an electric heating element. The total weight of this anti-icing system for a trawler is about 405 kg [1].

The total thickness of the heating elements of the system, developed by Napier and Son Ltd., is 1.5-1.7 mm and the mass is 1.46-1.70 kg/m^2. The insulation is designed for a temperature up to 100°C. Each square centimetre of the heaters consumes up to 6.2 W of elecric power. A thermostatic switch can be used to regulate the amount of thermal energy generated by the heating elements [1].

The specific consumption of power in electrothermal anti-icing systems reaches 1600 W/m^2. The limited power resources of the icing-up objects makes it possible to protect only small areas — major elements of the structure. The high cost, complicated design and great mass (up to 1% of the aeroplane's take-off weight) are also disadvantages of thermal anti-icing systems.

The electrothermal method is currently used on a wide scale to combat icing of aerial lines and aerial power lines under the regime of preventive heating to melt the forming ice by dc (Author's Certificate 649,078 USSR), and ac (Author's Certificate 705,583) USSR) [25].

THERMAL DESTRUCTION OF ICE

2.6. Inventions for Thermal Destruction of Ice

Nos	Country	IPC,NPC	Patent No.	Filed	Year of Publication	Filed by
1	USSR	84a,5/01	127,191	10 June 1959	1960	V.S. Khristoforov et al.

For wave reduction by charging air and water into upper layers of water body. To raise the effectiveness, pulse air charging alternates with water supply in the same direction.

| 2 | USSR | 84a,5/01 | 127,192 | 10 June 1959 | 1960 | V.S. Khristoforov et al. |

For wave reduction. Comprises an air line and valves. To alternate the supply of air and water the valves are in the form of a body with a cover and a bottom, in which there are apertures for water that close automatically when compressed air enters the body. The body cavity, connected to an air line, comprises three pipes. The internal and middle ones pass through the body cover and are used to discharge air or water from the cavity. The external pipe, which does not reach the cover, is a waterlock.

| 3 | USSR | 84a,5/01 | 127,193 | 10 June 1959 | 1960 | V.S. Khristoforov, I.V. Zagryadsky |

Hydraulic valve for a pneumatic breakwater with a supply of air. To elininate mechanical members, it is in the form of a protector cap with a perforated surface and a cylindrical body with apertures in the bottom. The body cavity, connected to the air line, comprises two pipes arranged concentrically. The internal pipe connects the body and cap cavities; the external pipe, connected to the bottom of the body, forms a waterlock.

| 4 | USSR | B63B,59; $65a^1$,10 | 195,910 | 6 Dec. 1965 | 1967 | V.S. Pechkovsky, P.M. Vakhmistrov |

For removing ice from the underwater part of a ship. A hollow rod with nozzles to supply steam. The rod is extensible and is fastened rotatably around a horizontal axis on a frame mounted on a portable support.

| 5 | USSR | B63B,59; $65a^1$,19 | 209,224 | 16 July 1968 | 1968 | V.S. Pechkovsky |

To prevent icing of the underwater part of a ship's hull. Comprises a flow generator with an electric motor and a steam line. The suction part of the flow generator comprises a closed circular pipe with steam outlets to raise the effectiveness of the thermal effect of the flow generator high-pressure jet on the iced part of the ship's hull.

| 6 | USSR | E02B; 65a^1,3; 84a,15/02 | 242,691 | 7 Mar. 1968 | 1969 | V.V. Salakhin, L.V. Ivanov |

For ice melting. Comprises a propeller pump in the extension piece with a straightening apparatus and guarding screen. A steam-water ejector with a diffuser directed toward the screw hub is assembled in the pump suction cavity to intensify ice melting.

| 7 | USSR | B63c,1/02 65B,6 | 266,608 | 6 Feb. 1969 | 1970 | L.V. Ivanov et al. |

For removing broken ice from under a ship's bottom. Comprises a pump with a base. To increase the effectiveness and automation of removing broken ice from under the bottom of a ship mounted on docking blocks, the pump's base is in the form of brackets with bearings on the dock's pontoon face. The pump's base is mounted rotatably on brackets by means of bearings in the diametral plane of the dock. The pump's centre of gravity is displaced from the horizontal axis of rotation toward the extension piece with an air cavity. The pump's turning catch is fixed near the pump's engine on brackets. The pump's face on the side towards the engine is connected with the dock pontoon by means of a cable, regulating the turning angle.

| 8 | USSR | B63B,59/00; 65a^1,19 | 280,250 | 24 Mar. 1969 | 1970 | N.F. Buyanov, N.V. Muzalevsky |

To prevent icing of sea-going ships. Comprises a source for heating the heating medium supplied to the parts of the ship subjected to icing. To increase effectiveness and reliability, it is in the form of a stationary system, comprising a trunk pipeline with branching pipes having cuts and sprayers, whence the heating medium discharges over the surfaces subjected to icing. The heating medium is sea water, which is supplied through two lines. One is connected to the cooling system of the main engine, the other to the sea water preheating system.

| 9 | USSR | E02B,15 | 279,446 | 7 Mar. 1968 | 1970 | I.A. Tuv et al. |

For making lanes in water bodies by supplying a heating medium and a gaseous agent. To intensify bubbling, the heating medium is first mixed with a gaseous agent and the mixture is discharged into the water body.

| 10 | USSR | B63B, 59/00 | 300,378 | 20 Sept. 1969 | 1971 | N.F. Buyanov, N.V. Muzalevsky |

For preventing icing of sea-going ships by washing the open surfaces subject to icing with a heating medium, such as preheated water. To increase the effectiveness, a continuous and controlled flow of heating medium is created. This flow is mixed with a flow of sea water splashes, flowing down the latter surfaces by the board. Sea water preheating is controlled so that the temperature of the water flowing down the washed surfaces is above the temperaure of the sea surface layer.

| 11 | USSR | B63c, 1/08 | 312,786 | 13 Mar. 1970 | 1971 | L.V. Ivanov, E.S. Vinogradov |

For removing broken ice from under a ship's bottom. Comprises a pump with a base. To increase effectiveness, the ship is put in a dry dock. The pump's base is arranged on the dock bottom at the caisson gate and includes a linkage comprising a rocker arm with a counterweight and a tie-rod, rotatably secured to the base and coupled to the pump.

| 12 | USSR | E21c,21/00 | 350,945 | 10 May 1970 | 1972 | V.A. Morev |

For electrothermal drilling of holes in ice with the formation of a core. Comprises a hollow circular body, whose cavity comprises an aligned row of copper rings with a heating element in between. To increase efficiency, the body is in the form of a cone with a copper ring extending outside the face.

| 13 | USSR | B63B,35 | 359,192 | 25 Apr. 1971 | 1972 | S.I. Yevdokimov et al. |

For maintaining navigational capability of ships in ice by washing the outside surface of the hull with the aid of nozzles arranged port and starboard. The water flow is created by a mixture of hot gases, for example steam and air, supplied to the contact area of the hull with the ice.

| 14 | USSR | B63B,59 | 360,268 | 17 July 1969 | 1972 | A.I. Borovsky et al. |

Steam-contact member. Designed to free a ship's hull of the ice cup in winter. To ensure continuous contact of the steam with the ice along the line of the working edges of the member, the holes are divided by transverse partitions and are restricted by lateral walls, forming nozzles of square section along the pipe.

| 15 | USSR | F25c,5 | 361,365 | 1 Dec. 1969 | 1973 | F.V. Peganov |

For making a lane in ice cover. Comprises a cylindrical body containing a heat source. To maintain the lanes ice-free, the body is in the form of a hollow, buoyant, hermetic ring with covers.

| 16 | USSR | E02B,15 | 367,212 | 18 Aug. 1969 | 1973 | N.V. Krapivko, A.G. Badyunin |

For melting ice. Comprises a propeller pump with an extension piece and straightening apparatus arranged in the zone of pump discharge, and a heater. The heater is built in to the straightening apparatus

17	USSR	B63B,35	382,544	2 Dec. 1971	1973	S.I. Yevdokimov, V.S. Kashtelyan

For maintaining navigational capability of ships in ice. Nozzles supplying a mixture of hot gas from the ship's propelling plant are arranged on the sides of the ship at the level of hull contact with the ice. The distinguishing feature is the jet apparatus connecting the nozzles with the propelling plant, sucking in air from the atmosphere. The connecting pipelines of the nozzles are equipped with shutoff and control valving.

18	USSR	E02B,15/00	394,494	30 Sept. 1970	1973	I.V. Bazilevsky

For creating a liquid current in a water body. Comprises perforated pipelines to supply compressed air. The perforated pipes are arranged at an angle to one another to create a directional current in the water body.

19	USSR	E21c,37	399,600$^+$	16 Feb. 1972	1973	B.B. Kudryashov et al.

Electric drill for drilling holes in ice. A flange is fastened on the face of a circular electric heating element to improve the supply of heat to the hole, and the branch pipes removing meltwater are arranged outside the core barrel. The open ends of the branch pipes are fastened in the flange and face the hole.

20	USSR	B63B,35	404,698	26 Apr. 1971	1973	Emile Mayer (FRG)

For increasing the navigational capability of an icebreaker in ice. Comprises an icebreaking unit mounted on a platform in the icebreaker's prow under the ice field. The width of the unit exceeds the width of the icebreaker. The icebreaking unit comprises turbojet engines arranged across the width of the platform, whence a reaction jet is directed at the ice cover at an angle towards the stern.

21	USSR	E21c,21	439,601	12 July 1972	1975	V.A. Morev

For electrothermal drilling of holes in ice. Comprises a body with core extractors and a crown bit, a piston arranged in a body with canals that connect the cavities above and beneath the piston.

22	USSR	E21B,7	446,617	8 May 1973	1974	A.P. Dmitriyev et al.

For drilling ice. Comprises a heat generator with nozzles, a pump and a heat exchanger. The heat exchanger is of conical shape to increase the drilling rate.

23	USSR	E02B, 15/02	499,379	21 July 1972	1976	N.D. Andreyev, L.V. Ivanov

For maintaining parts of watercourse area ice-free. Comprises a water lift and an anchoring system. To reduce power consumption during operation, the water lift is in the form of a pipe whose bottom end is connected to the anchor, and is equipped with a float in the form of a rudder-stabilizer.

| 24 | USSR | E21c,21 | 564,419 | 9 Mar. 1976 | 1977 | B.B. Kudryashov et al. |

For drilling holes in ice with coring. Comprises a core barrel with core-holding teeth, equipped at the face with a hollow heating ring, electric steam generator, and water receiver with a pump. The space in the heating ring is divided by radial partitions into isolated cavities, some of which are connected to the steam supply pipes, others to condesate suction pipes, which are connected to the water suction pipes.

| 25 | USSR | E21B,7 | 581,227 | 10 June 1976 | 1977 | A.A. Kapustin et al. |

For drilling ice. Analogous to USSR Author's Certificate 446,617. To intensify ice drilling by reducing heat losses to the atmosphere through the upper cover of the heat exchanger, the latter is divided by a partition into two chambers. The upper chamber communicates with a compressed-air trunk line and the combustion chamber of the heat generator.

| 26 | USSR | E21B,7 | 581,228 | 14 June 1976 | 1977 | A.P. Dmitriyev et al. |

For drilling ice. Analogous to USSR Author's Certificate 446,617. Comprises branch pipes with a variable cross-section along the length, narrowing from the inlet to the central part, and widening to the outlet.

| 27 | USSR | E21c,21; F25c,5 | 585,285 | 2 Aug. 1976 | 1977 | A.F. Strelenko, V.P. Shmatko |

For drilling holes in frozen soil or in ice. Comprises a heat source and a core receiver with double walls and sharpened working tip which is equipped with a heating element and a cooling element. The heating and cooling elements are in the form of a layer of porous material saturated with vapourizing fluid on the inside of the pipe wall, and the heat source is arranged in the tail part.

| 28 | USSR | B63c,1 | 647,179 | 15 Aug. 1977 | 1979 | L.V. Ivanov, N.D. Andreyev, E.S. Vinogradov |

For protection of a floating dock against broken drift ice. Comprises a source of compressed gas and a perforated pipeline connected to it. The pipeline is suspended by the contour to the dock floor. The pipeline's suspensions are in the form of a solid vertical wall, and the perforated pipeline is mounted on the wall side external to the dock.

| 29 | USSR | H02g,7 | 649,178 | 27 Nov. 1974 | 1979 | V.V. Burgsdorff et al. |

For melting slush by direct current. A filter of industrial frequency current is connected in parallel to the terminals of a constant voltage rectifier to restrict the overvoltage on the rectifier which arises in the slush melting circuit on energizing and de-energizing of the latter circuit.

| 30 | USSR | H02g,7 | 705,583 | 1 Aug. 1977 | 1979 | G.V. Shinkarenko et al. |

For melting slush on AC power lines. Comprises a current converter and a regulator. To increase reliability, it is equipped with a biasing pulse generator, whose output is connected to the current regulator input, and the input is connected via the OR circuit to the outputs of the current regulator, feeding a signal to the control electrodes of the rectifying groups of valves. The converter comprises a circular circuit.

| 31 | USSR | E21B, 21/00 | 794,178 | 19 Nov. 1978 | 1981 | V.A. Morev, V.A. Pukhov, A.G. Nikiforov |

For electrothermal drilling of a hole in ice. Comprises a body, a crown bit, a core receiver, a piston, and a core receiver cavity above the piston filled with antifreeze. To increase the effectiveness of drilling, it is equipped with an antifreeze tank with a piston and a rod, as well as discharge and suction conduits.

| 32 | USSR | B63j,5/00 | 844,475 | 20 Aug. 1979 | 1981 | V.A. Morev |

Ship-borne means of ice destruction. Comprises a hollow body of wedge-like sections, an inlet and an outlet pipe. To raise the effectiveness of operation, a series of through longitudinal conduits are comprised in the bottom part of the body in parallel to one another.

| 33 | Great Britain | 113 K | 805,789 | 4 Feb. 1957 | 1958 | Pneumatic Breakwaters Limited |

Improvements in artificial breakwaters. The breakwater comprises a container which is so designed that by a siphon effect an intermittant discharge of air or other gas takes place in the form of large bubbles which inhibit wave motion in the surrounding sea water. A supply of air from a compressor is conveyed to the sea bottom through a pipe with adjustable valves. The pipe rests on the sea bottom.

| 34 | Great Britain | 113 n | 894,878 | 1 Feb. 1960 | 1962 | Alfred Jönsson |

THERMAL DESTRUCTION OF ICE

A method of heating ships' hulls, and equipment for carrying it out. Comprises a screening device extending down into the water, and a heating device insertable between the screening device and the ship's hull for heating the water within the screening device, and a pipe system for supplying warm water, the pipes in the system having upwardly directed perforations. The screening device comprising two gas-inflatable superimposed tarpaulin canvases with a space in between. The tarpaulin canvas facing the ship's hull is shaped in the form of successive waves while the other tarpaulin canvas is plane.

35	Great Britain	B7MK, B63b	967,543	30 Oct. 1959	1964	Aero-Hydraulics Limited

Improvements in wave reduction, de-icing and destratification. A method of reducing sea waves and swell, or destratifying liquid, or de-icing, or inhibiting the formation of ice on bodies of water, consisting in feeding bubbles of air or gas into the lower end of a conduit in such a manner that they ascend toward the surface of the liquid, carrying with them a substantial body of liquid through such conduit, and ejecting the bubbles and ascending liquid at the head of the conduit below the surface of the liquid, whereby turbulence is generated in the ambient liquid, and large bubbles are fed into the lower end of the conduit by emission through an ejector.

36	Great Britain	B64D,15; B7W	1,308,896	18 Jan. 1971	1973	Goodyear Tire and Rubber Company

De-icing device. A de-icing device for aircraft or ships, including two layers of flexible material and a system of conduits. Compressed gas is supplied through the conduits arranged on each side of the member being de-iced. The conduits communicate with a manifold conduit. Air heating.

37	Great Britain	B63B,35; B7A	1,315,834	6 Aug. 1971	1973	John V. Buckland

To assist the passage of ships through ice. A vessel is provided with some means to apply heat to surrounding sea ice to cause it to melt and so free the vessel. The means comprises a series of apertures extending through the hull of the vessel at or below the waterline and connected to a source of either steam or hot gas within the vessel. The heat is principally or entirely produced by the vessel's means of propulsion. The nozzle outlets may be arranged at the ends of telescopic arms. Heat may also be generated by electric or atomic means operating together with an acoustic radiator.

38	Great Britain	E21c,21; E1F	1,315,921	4 Jan. 1972	1973	France, Armed Forces

For forming a hole in ice. Comprises a body member with a ram extending from it and having associated with it generating means for generating hot gases and directing means for directing hot gases circumjacent to the ram.

| 39 | Great Britain | B64d,15; B7W | 1,351,158 | 20 Aug. 1970 | 1974 | B.F. Goodrich Company |

Lead for aircraft propeller electric de-icer system. An electric de-icer system having a boot containing a resistance-type electric heater mounted on a propeller blade, a lead strap containing wires. The heater is connected to a power source on the main body of an aircraft through a slip ring mounted on the propeller assembly and brushes mounted on the aircraft.

| 40 | Canada | 255-10 | 925,849 | 19 July 1969 | 1973 | E. Horback |

For drilling holes in ice. The drilling unit comprises a hollow drill pipe with a circular melting head at the bottom end. Insulation is inserted between the drill pipe and melting head. Cooling pipes arranged above the melting head are for freezing the walls of the drilled hole.

| 41 | Canada | 114-11 | 964,527 | 19 Dec. 1972 | 1975 | Friedrich J. Legerer |

Ice-breaking apparatus. An apparatus for forming a slit in an ice mass in the path of travel of a vessel. A flexible hose is dragged in contact with the ice and jets of a heated medium are directed from the hose against the contacted area of the ice.

| 42 | USA | 61-1 | 3,083,538 | 6 Oct. 1958 | 1963 | George E. Gross |

For maintaining a body of water free of ice. Comprises a circular housing, a submersible motor unit with an output shaft, means for mounting the housing at a substantial distance below the surface of the body of water. The housing has a water outlet. The apparatus moves water from the depth of the water body to the surface as a stream free of mechanical restraint and without substantial mixing of the moved water with the surrounding body of water. The moved water spreads out as a protective layer on the surface of the body of water.

| 43 | USA | 61-1 | 3,109,288 | 19 Jan. 1960 | 1963 | Perma-Pier, Inc. |

Oscillating and aerating ice and water control. Air is moved from above the surface into the body of water. The air is divided into fine bubbles that remain in suspension. The mixture of water and bubbles sets up an aerated column of reduced density. The column will rise to the surface of the body of water in a defined area.

| 44 | USA | 61-6 | 3,148,509 | 24 Oct. 1960 | 1964 | Pneumatic Breakwaters Limited |

Wave reduction, de-icing and destratification apparatus. A liquid-circulating apparatus for moving a liquid upwards, comprising a conduit with a lower liquid intake opening and an upper discharge opening, and an intermittent gas bubble generator with an upwardly directed bubble dscharge port.

| 45 | USA | 61-1 | 3,170,229 | 27 Apr. 1962 | 1965 | John H.O. Clarke |

For prevention of ice damage to boats, piers and the like. Comprises a metallic heat-conducting element and watertight heat insulating material around the outer face of the heat-conducting element. The heat-insulating material has continuous closing upper and lower terminal edges with the vertical dimensions between them being greater than the thickness of ice resulting from natural icing conditions. A heat-conducting element is supported in the water with its upper edge at the surface of the water in its ice-free condition and its lower strata of warm water beneath a layer of ice resulting from natural icing conditions.

| 46 | USA | 261-64 | 3,193,260 | 13 Mar. 1961 | 1965 | C.M. Lamb |

For aerating and eliminating ice on water. An apparatus comprising a casing member having an outer peripheral member, ribs extending downwards and inclined toward each other, a lower end member joining the lower ends of the ribs and top frame member joined to the outer peripheral member, a top wall mounted on the top frame members and a lower wall mounted on the outer sides of the ribs, air discharge means for discharging air bubbles for upward flow to cause upward and outward movement of water about the casing member, means connecting the air discharge means to the casing member, and a supply of air under pressure to the air discharge means.

| 47 | USA | E02b,15; 61-1 | 3,477,233 | 7 Mar. 1966 | 1969 | P.F. Andersen |

Wave machine installation. A machine for making waves on the free surface of a body of liquid by a buoyant member which is free to make vertical movements with the liquid level as the latter changes. The buoyant member is of elongated shape. A drive is connected to the buoyant member to cause its rotation about a rotary axis. Suitable for freeing channels and harbours of ice.

| 48 | USA | E02,b,15; 61-102 | 3,759,046 | 23 Mar. 1972 | 1973 | Global Marine, Inc. |

Movement of marine structures in saline ice. Apparatus for facilitating relative motion between an ice sheet overlying a body of salt water and a structure disposed in the water and extending through the water surface, the method comprising the steps of applying thermal energy from the structure to ice adjacent to the structure in quantities and at rates sufficient to heat that ice to a temperature which is within the melting point of the ice, and mechanically reducing the weakened ice pieces sufficiently small to pass readily around the structure.

| 49 | USA | E02b,3; 61-1R | 3,768,264 | 3 July 1972 | 1973 | The Dow Chemical Company |

For suppressing the formation of ice in natural or man-made bodies of water. The method involves the use of heat exchangers for direct replacement of heat loss to the atmosphere from the water during the winter season. The method is particularly useful in areas where ice-breakers or the like cannot be operated efficiently.

| 50 | USA | B64B,15; 219-202 | 3,800,121 | 22 Mar. 1972 | 1972 | Michael Gordon Ellis Dean, Brian James Saunders |

Electric heating apparatus for reducing or preventing the formation of ice on aircraft parts. Comprises a metallic layer forming an electrical resistance heating element. The metallic layer is secured to a conducting layer attached to the non-metallic external surface.

| 51 | USA | B63B,35; 114-5D | 3,837,311 | 5 Oct. 1972 | 1974 | Sun Oil Company |

Apparatus for melting ice. A surface-effect drilling vehicle with ice-melting apparatus has a hollow hull lined with a fibrous material and a low-vapour-pressure liquid therein. Heat is applied to the hull's interior wall and vaporizes the liquid, causing it to transfer by diffusion towards the exterior wall of the hull where it loses its heat to the ice and condenses. By the capillary action of the fibrous material the condensed liquid returns to the heated hull interior wall, and the process is repeated.

| 52 | USA | B63B,35; 114-40 | 4,075,964 | 29 Aug. 1975 | 1978 | Global Marine, Inc. |

Ice-melting system. A heat transfer apparatus comprising a series of elongated grooves in the exterior surface and located next to the portion of ice to be melted. A portion of the grooves extends upwards to a location above the surface of solid ice. Heat can be directed outward from the grooved surface to a portion of the layer located next to the grooved surface. The grooves provide areas of heat concentration, affecting the layer to be melted. The meltwater can drain away.

| 53 | USA | B63B,35; 114-40 | 4,117,794 | 11 Apr. 1977 | 1978 | Global Marine, Inc. |

Ice-melting system and method. In the path of movement of ice is disposed a portion of an exterior heat conducting surface with several elongated grooves adjoining the portion of ice to be melted. Heat is directed continuously from the exterior surface to the portion of the ice layer to be melted. The grooves provide areas of heat concentration and act as conduits for removing melted ice.

THERMAL DESTRUCTION OF ICE 175

54   France   E02b   1,211,708   19 May, 1958   1969   George E. Gross

For ice prevention on a water surface.

55   France   E02b   1,237,407   6 Oct. 1958   1960   George E. Gross
                        Analog of
                        US patent
                        3,083,538

For maintaining a body of water free of ice. Comprises a circular housing having a water outlet and a water inlet. The self-contained apparatus moves water from the depth of the water body to the surface as a stream free of mechanical restraint and without mixing of the moved water with the surrounding body of water; the moved water spreads out as a protective layer on the surface of the water body.

56   France   E21c,21   2,166,393   7 Jan. 1971   1973   Jacques C.
                                                            Delgendre

For drilling ice with a hot gas generator. Thick ice layers are drilled automatically and rapidly. Comprises a combustion chamber to generate hot gas, with outlet nozzles heating the drilling zone around the tip. A buoyant capacity is used for levelling.

57   France   64d,15   2,202,810   16 Oct. 1972   1974   Pneumatic Caout-
                                                            chouc Manufacture
                                                            et Plastique
                                                            Kleber-Colombes

Thermoelectric de-icer for blades, etc. Comprises one thin heating layer of heat-resistant elastomer, maximum thickness 0.5 mm.

58   FRG   84a,5/01 1,075,507   13 July 1957   1960   Pneumatic Break
                                                            waters Limited

Pneumatic breakwaters.

59   FRG   84a,   1,166,706   6 Oct. 1958   1964   George E. Gross
              15/02;   Analogue
              E02b   of US
                        patent
                        3,083,538

Pump for maintaining a body of water free of ice.

60   Finland E02b,   32,261   19 May 1958   1978   George E. Gross
              84a,6/01

For ice prevention on water surface.

61	Japan	B63B,59; 84a,6/01	53-30,255	24 Oct. 1973	1978	K.K. Inoue dzyapakkusu kenkyudze

For destruction of ice cover. A wire connected to terminals is arranged on the surface to be freed of ice. A current pulse is supplied through the wire, which instantaneously melts, creating local pressure which destroys the ice cover.

62	Japan	B63B, 35/12	55-46,915	10 Apr. 1976	1981	Mitsui dzosen K.K.

Icebreaker-mounted apparatus for cutting ice. A slot with nozzle is in the prow of the icebreaker, which contacts the floes directly. The end of each nozzle is surrounded by electric heating elements. Each nozzle is connected to a control valve by lines that are heated by heating coils. The valve is connected by a pipe to a high-pressure compressor. The pipe is connected through by-pass valves with the line for rapid removal of water from the compressor.

Chapter 3

CHEMICAL DESTRUCTION OF ICE

The destruction of ice by means of chemical substances is based on the property of some substances to form mixtures with ice that have a lower melting temperature than their components. This is called the eutectic temperature, and the concentration of the eutectic solution is dependent on the properties of the chemical substance. Ice melts in an eutectic solution up to the maximal concentration of the brine corresponding to the given temperature. The character and degree of ice destruction depend on the substance, size of particles, spraying norms, temperature and structure of the ice. Ice melts in a uniform layer from the top downwards under the effect of powdered chemical substances. Individual lumps of chemical substances penetrate into the ice, forming twisting canals with strong partitions. The ice strength is reduced because of violation of its solidity.

At -5°C 1 g of potassium bicarbonate can melt 50 g of ice, sodium fluoride 33 g, and sodium sulphide 21 g. It is advisable to use other salts at lower temperatures: thus, ammonium, sodium, and potassium chlorides are used at -6 to -20°C [15, 136, 180] (Table 3.1).

A brine site forms at the point of contact of the chemical substance with the ice, and this site deepens as the temperature rises. A maximal amount of ice is melted at sharp temperature rises. Sodium, potassium, and ammonium chlorides melt ice mainly vertically, while magnesium chloride acts destructively sideways. (Table 3.2).

A dependence exists between the consumption of chemical substances on the ice temperature and thickness.

When destroying 10 cm thick ice on an area of 1 ha at -10°C, it is necessary to spend 3.5 t of salt in lumps of at least 10 g. To destroy weak river preflood ice of the same thickness and on a same area, 0.5 t of small-grained salt is enough.

Table 3.1
Volume of melted ice per 1 kg of salt at various temperatures (cm^3)

Chemical substance	Eutectic temperature (°C)	Temperature (°C)			
		-5	-10	-15	-20
Calcium chloride	-55	10.8	6.4	5.0	4.1
Sodium chloride	-21.2	12.2	6.7	4.7	3.7
Ammonium chloride	-15.8	14.0	7.1	4.8	-
Potassium chloride	-11.1	10.3	4.7	-	-
Sodium sulphide	-10.0	21.0	10.6	-	-
Sodium fluoride	-5.6	33.0	-	-	-
Potassium bicarbonate	-5.4	59.0	-	-	-

Table 3.2
Change of depth of ice destruction by chlorides at a concentration of 1 g/cm^2 at temperature drop (cm) [15]

Chemical substance	Temperature (°C)								
	-11.7	-10.2	-8.5	-6.5	-5.5	-4.5	-3.0	-4.5	-3.0
	Freshwater ice							Saline ice	
Sodium chloride	6.0	7.0	7.1	7.1	7.2	7.7	8.0	8.8	25.0
Potassium chloride	0.0	0.0	5.3	8.0	9.0	9.0	9.0	8.0	9.0
Ammonium chloride	4.0	6.4	6.5	6.7	7.2	7.6	8.2	8.6	25.0
Magnesium chloride	1.6	1.8	2.2	2.2	2.4	2.4	2.6	2.6	2.8

The consumption of chemicals in warm weather at a stable above-zero mean daily air temperature can be reduced by 50%. The consumption of the chemicals is usually 7-10 times less than the dissolved ice mass in the case of overall ice melting (without leaving ice bridges).

The width of a strip sprayed with chemicals from a vehicle is 0.3-0.4 m; when the chemicals are sprayed from an aeroplane it is 2.5-3.5 m. The use of aircraft has made it possible to increase the degree of mechanization, but the consumption of chemical substances increases.

The experience of using aviation-dispersed chemical methods for removing ice from the approaches to a port has demonstrated that the destruction of ice occurs 15-20 days earlier, with the consumption of 50-100 t of rock salt per hectare, and the prime cost is about 2% of the expenditure for ice destruction by means of icebreakers [6].

It is advisable to use the chemical method on a limited area for local destruction of ice. It is usually considered as an auxiliary method and is used together with ice-cutting machines on river stretches with a greater ice thickness (when the thickness exceeds that of the milling cutter), or containing a large number of logs and other solid objects, as well as at intersections of ice-cutting routes.

The advantage of the method is the rapid effect of the chemicals on ice. Lumps of salt with a size of 2-2.5 mm to 4-4.5 cm, scattered on ice under natural conditions, can penetrate it to a depth of 20 — 70 cm in 24 hours if the ice has a crystalline structure and the air temperatures are above zero.

The high cost of the materials and reduction of effectiveness owing to solubility of the salts in the presence of water and snow on the ice surface, as well as water interlayers inside the ice, are disadvantages of this method.

Sleet and icy snow on asphalt-concrete roadway coverings demand increasing attention. They are eliminated by means of mechanical, thermophysical, chemical methods, and combinations of these. The blades of bulldozers and graders are of little use, because mechanical impacts usually destroy the roadway covering.

The chemical method is the most effective and economical one for eliminating slush on roads, but the frictional method is still used simultaneously to overcome slipperiness in winter, when sand, small gravel, fuel slag, or comminuted stone with a 5 mm maximal particle size is scattered at a rate of 200-300 g/m^2.

Among other methods, rock salt is used with inhibitors to prevent corrosion, on the take-off decks of US aircraft carriers. Rock salt is sometimes replaced by a lithium chloride mixture. The latter mixture is effective on ice formed from snow, but it is ineffective when ice forms as a result of sea water splashes, because in this case it is washed off. A lithium chloride mixture can be used if the ice thickness is slightly more than 1 cm [1].

Chemical substances are used in liquid and solid forms to eliminate slush. Calcium chloride solutions of 32% or 35% concentration, sodium chloride solutions of 20% concentration, and more concentrated natural underground brines, whose cost is from 6 to 20 kopecks per cubic metre [135], stratal water of oil deposits, brines of saline lakes, and chemical industry wastes are used on a wide scale. Solutions of urea, ethylene glycol, and other substances such as a reagent mixture of 80-90% calcium chloride with 20-10% calcium nitrate nitrite are used with success on automobile roads [28, 39].

Solid crystalline substances, such as sodium chloride, phosphated calcium chloride, as well as a mixture of 12% calcium chloride and 88% sodium chloride are used with favourable results on the Moscow—Leningrad highway [39, 68].

The best results have been gained by using flaked calcium, which melts ice at an air temperature of -30°C [39].

An instruction for combatting slush on automobile roads, elaborated at the Road Research and Development Institute, has been in effect since 1 January 1975 [54]. This is an accepted standard.

Experiments on determining the reduction of ice strength by means of surfactants have demonstrated that the strength is reduced by 20% when an aqueous solution of the chemical is used [36, 130], and the application of a surfactant film on freshwater ice reduces the strength of the latter by 30% [66]. According to recent data, application of the Rebinder adsorption-surface effect can reduce the ice strength by 80%.

An urgent objective is to develop a constructive method of eliminating slush on roads, by introducing new roadway coverings.

One of the latter (Author's Certificate 607,868 USSR) comprises an upper wear-resistant layer of an organic binding substance and a filler, which is rubber fines 1-10 mm in size at a level of 0.63-0.72 m^3/m^2.

Another (Author's Certificate 280,517 USSR) is a composition for covering air fields, roads, and other structures, based on waterglass, blast-furnace granulated slag, and sand. Clay and an aqueous solution of alkali metal are added to reduce adhesion of the ice to the surface of the covering.

A powdered anti-slush reagent causes ice melting, and the forming solution penetrates into the layer under the ice, contributing to rapid flaking of the ice from the surface of structures.

The essence of physico-chemical methods of combatting icing consists in creating an intermediate layer of some special substance between the ice and the protected surface. The substance should either reduce adhesion of the ice, or lower the temperature of sea water freezing on a ship's surface. It is desirable that the coating should combine both of these properties. Research has shown that cryophobic organo-silicate and organo-silicon coatings and saline antifreeze solutions can be used with success for the protection of metallic surfaces, bridges, roofing, and ferries against icing [66].

The use of anti-icing coatings of organo-silicate materials [74], enamel EP-51-62 [120], polycarbonate or polysiloxane block copolymers [203], ethylene or teflon sprayed on the surfaces of structures [26], and technical vaseline [1, 26] is currently recommended.

Water-resistant ropes should be treated with solid vaseline (petrolatum). The rubbing parts of blocks should be coated with grease, and the external parts with petrolatum. The process of chemical treatment should be repeated periodically. The time interval between coating cycles depends on the weather.

Publications in various countries report the development of a special anti-icing coating X-3057 for the surfaces of ships, reducing by 70% the operations of chipping off the ice and preventing icing by its water-repellent properties. However, the lifetime of this material is short, about two weeks, owing to its sensitivity to solar radiation and weak adhesion to the coated surfaces.

Teflon, which is manufactured as a thin film, is characterized by good anti-icing properties. It can be wrapped around masting, rigging, antennas, etc. Ice is removed easily from Teflon, and from flexible elements such as antennas, the ice is removed without human interference by the effect of wind and vibration.

The organo-silicon-epoxy coating EP-51-62 comprises two components: a corrosion-resistant intermediate layer of pigmented lacquer based on epoxy resin ED-5 with a hardener, and an oriented layer of an organo-silicon polymer, chemically associated with the corrosion-resisting intermediate layer, comprising an organo-silicon fluid GKZ-94 with a hardener. This coating, grade G, has an adhesion strength of 0.981 Pa.

The effectiveness of the coating depends on the condition of the protected surface and the hydrometeorological conditions. In some cases the coating has operated successfully for six months, in other cases for less than one month [1, 120, 123].

Organic antifreezes that have demonstrated good results under laboratory conditions are cryophobic preparations on an organo-mineral base. They are solid substances with a melting temperature of 100-150°C. Ice melts across the area of adhesion with the protected surface and is easily moved and flaked off [136].

Fluoroplastic covers are effective in protecting rigging. The ice is easily displaced across the cover and disintegrates on impact [134].

Soviet industry manufactures hydrophobic organo-silicon polymers in the form of fluids, resins, lacquers, etc. The adhesion of ice with these coatings is two times less than with standard coatings, and three or four times less than with metal.

The substitution of metal structures by plastic and glass-reinforced plastic is of certain interest. It is advantageous in the case of icing to use paraphyllic cables, where ice freezes slowly on the smooth surface. A paraphyllic cable is made of parallel fibers of high strength, which are pressed and enclosed in a flexible alkathene casing. They are manufactured of nylon, terylene, polypropylene, or their compounds.

The use of plastics based on polyurethane (pat. 225,249 France), sponge materials, polyethylene foam plastic [120], porous titanium, and polyethylene film (USSR State Standard 10354-64, grade A) has been suggested [26]. The protected surface is coated with one of these hydrophobic materials, greatly reducing the adhesion of the ice and retarding ice build-up under certain hydrometeorological conditions. Icing can also be prevented by using porous titanium, through which air or oil is pumped under pressure.

The surface of solid hydrophobic polymeric coatings is characterized by low free energy and low ice adhesion — 10 times less than in the case of ordinary paint coatings.

Some of these results have been introduced in Leningrad to eliminate the icing of roofs. Practice has demonstrated that the treatment of roofs with compact coatings, comprising a layer of epoxy and organo-silicon polymer (epoxy enamels EP-51-62 and underpaint putty EP-00-01) reduce almost 10 times the adhesion of the ice to the surface [22].

Adhesion of the ice to the protected surface is reduced by spraying a mixture of polyether and isocyanate with the addition of silicon oil Pat. 4135/73 Japan), or fluoric resin (pat. 4134/73 Japan), as well as by applying coatings of urethane resins (application 41316/73 Japan).

Solid saline coatings and solutions are used in aviation, on roads and runways and are not used in practice for protection against splash icing. Saline coatings are of low effectiveness and are not economical because of the high consumption of the substance and destruction of the coatings under the effect of atmospheric conditions. They also cause intensive corrosion of the protected structure.

Fluid anti-icers ensure a continuous supply of fluid and its uniform distribution over the protected surface. Chemical, mechanical and thermal effects preventing icing occur when atmospheric moisture precipitates on a surface coated with a fluid anti-icing substance. Polar organic fluids with a low freezing temperature and easily dissolved in water are used as working fluids: for example, ethyl, propyl and butyl alcohols, and glycols. Fluid anti-icing systems are effective only for the prevention of icing [136].

The consumption of the working fluid is up to 500 cm^3/min. The effectiveness of fluid anti-icing systems has increased with the application of porous metals, for example, titanium, bronze, etc., for manufacturing the distributing devices.

Vaselines are organic, organo-silicon, or fluorinated fluids, thickened by finely dispersed fillers. When applied to the surface of an iced object, they preserve a pasty condition at icing temperatures, and the strength of ice adhesion to the substrate is determined by the strength of the vaseline layer, which increases the chipping effectiveness two to three times.

It is advantageous to use vaseline systems to protect extended surfaces, when the application of energy methods is impossible, as well as to protect the antennas of radio relay systems.

Combined anti-icing systems are energy anti-icing systems of which the surfaces have a low free energy and low ice adhesion. This combination makes it possible to increase efficiency and reduce power consumption per unit area of protected surface. Combined physicochemical anti-icing systems comprise immiscible fluids, pastes, and embrittling surfactants.

The free surface energy of the ice is reduced and the brittleness of an ice block increases according to the Rebinder effect, as a result of interaction of the ice on the surface of an object with surfactants. A relatively slight reduction of the surface energy is enough for the manifestation of this effect [142]:

$$(\sigma_{F,rb})^2 = (\sigma_{F,D})^2 - (AE/C)(\gamma_o - \gamma_{rb})$$

where $\sigma_{F,rb}$ is the respective breaking stress of ice with a modified surface, $\sigma_{F,D}$ is the same for ice with a surface grown in vacuum, A is a constant numerical coefficient, E is Young's modulus, γ_o is the free surface energy of ice in vacuum, γ_{rb} is the same for modified surface, and C is the length of largest, most favourably oriented crack.

Organic alcohols, fatty acids, amines, and sulphonic acids are the most efficient surfactants, and perfluorinated analogues are still more effective.

The increased effectiveness of a combined anti-icing system with embrittling surfactants applied to the external surface of ice grown on a low-energy surface, as compared to the effectiveness of the same system for ice grown on a high-energy surface, can be demonstrated by the operation of the electric pulse anti-icing system (EPAS) (see §4.1).

Keeping all other parameters constant, the use of surfactants raises the effectiveness of ice destruction by EPAS on high-energy surfaces by 100%. The use of surfactant films in combination with low-energy surfaces raises by 200% the effectiveness of ice destruction by EPAS as compared to high-energy surfaces [142].

3.1. Inventions for Chemical Destruction of Ice

Nos	Country	IPC,NPC	Patent No.	Filed	Year of Publication	Filed by
1	USSR	E10c,7	280,517	10 Jan. 1969	1970	K.N. Andrushchak et al.

Composition for coverings of aerodromes, roads and the like. A composition for coatings to protect the latter against icing, based on waterglass, blast-furnace granulated slag, and sand. Clay and an aqueous solution of alkali metal are added to reduce adhesion of the ice to the covering surface.

| 2 | USSR | E01c,7 | 607,868 | 17 May 1976 | 1978 | V.S. Borovik et al. |

Covering for automobile road. Comprises a base and an upper layer of an organic binder and fine particles of a filler. To prevent slush and reduce tyre wear, the fine particles of the filler in the upper layer are made of rubber with a size of 1-10 mm. The consumption rate is $0.63-0.72 \text{ m}^3/100 \text{ m}^2$.

3   France   B63B,59;   2,252,249   23 Nov. 1973   1975   Goodyear Tire and
             C09K,3   Rubber Company

Transport vehicle with protective coating for protection of metal surfaces in contact with ice. The protective coating is made of plastic on the basis of 0.25 to 1.15 mm thick polyurethane.

4   Canada   255-10   969,168   27 Nov. 1972   1975   Frederick D. Ward

Thermal penetration of ice. A body for piercing ice, comprising a core, formed of a high-specific-gravity substantially water-inert solid material, namely iron, steel, or concrete. Depending on the ice thickness, the shell surrounding the core includes a chemcial which has a highly exothermic reaction with water, namely an alkali metal or its hybride. The reaction between this material and the ice generates sufficient heat to melt the adjacent ice.

5   Japan   B63B,3;   48-41313   17 Apr. 1969   1974   Nippon baruka
             84B3   coge K.K.

For prevention of icing. A mixture of polyether and isocyanate with added fluorinated resin is applied to the ship's hull.

6   Japan   B63B,3;   48-41315   21 Apr. 1969   1974   Nippon baruka
             84B3   coge K.K.

For prevention of icing. A mixture of polyether and isocyanate with added silicon oil is sprayed on.

7   Japan   B63B,3;   48-41316   12 Sept. 1969 1974   Nippon baruka
             84B3   coge K.K.

For prevention of ship's hull icing. A binding layer is applied to the ship's hull surface, which is successively coated with urethane, foam polyurethane, and urethane elastomer by spraying.

Chapter 4

ELECTROPHYSICAL DESTRUCTION OF ICE

The necessity for more effective destruction of rock in the extraction of useful minerals, and development of frozen soils for the construction of water-development works and in transport and civil construction beyond the Arctic Circle, have encouraged research and development work on utilizing the radio-frequency magnetic field, ultrasound, irradiation, lasers, and low-temperature plasma jets [115, 181]. Attempts have been made to apply some of these methods for the destruction of ice, but most have not gone beyond the stage of experiments on prototypes and it is therefore possible to discuss technical aspects of this method only to a first approximation.

4.1. Electrical Pulse Technique

The electric pulse method of ice destruction is based on the interaction of a magnetic pulse field generated by an inductor with eddy currents induced by this field in a metal structure whose surface is to be freed of ice. A substantial mechanical force develops as a result of this interaction, deforming the thin-walled structure; as a consequence, the ice layer cracks and flakes off [79].

An electric pulse anti-icing system (EPAS) has been developed in the USSR on the basis of this principle. EPAS comprises an electric pulse generator, supplied by the on-board power network, several dozen inductors — devices transforming electric pulses into mechanical ones — and distributing elements.

When the system is energized, the generator supplies short electric pulses with long pauses in between, accumulating energy during these pauses (Author's Certificate 201,086 USSR). The power in the pulse exceeds by a factor of 3600 the consumed average power of the system at a pulse duration of 500 ms, a 2 s pause between the pulses, and 90% efficiency of the accumulating block.

The programme commutator, opening the semiconductor keys in the given regime, supplies these pulses to each inductor in turn. The powerful pulse generates eddy currents in the thin-walled metal casing. Interaction of primary and induced currents develops forces causing deformation of the casing and of the ice layer frozen to it. The stress arising in the casing is below the limit of cyclic strength, and sufficient for destruction of the ice layer. Because of the elastic properties of the casing the deformation wave propagates from the spot of its formation across the entire zone protected by the inductor, destroying the frozen ice layer. The specific power consumed by the system is 25-50 W/m^2.

The experience of operating EPAS (Authors' Certificates 213,588 and 213,590 USSR) in modern aircraft has demonstrated the effectiveness and economy of this system. It is patented in France (pat. 1,562,244), Italy (pat. 843,087) and Norway (pat. 117,890).

EPAS operates most efficiently in combination with special elastic casings and surfaces with a low specific free superficial energy, as well as when ice is treated with surfactants.

Investigating the effectiveness R of pulse destruction of ice, the authors of [65] propose calculating it as a function of the physico-mechanical properties of the base, the ice, and EPAS paramaters:

$$R = L(F_{ad}^S/F_{ad})^m \cdot N(\sigma_s/\sigma)^P$$

where F_{ad} and F_{ad}^S are the forces consumed for destruction of the adhesive contact of anti-icing and standard coatings respectively, σ and σ_s are strength characteristics of the ice frozen on the anti-icing and standard coatings and L, N, m, P are weighing factors taking into account the contribution of the adhesion and strength members to the energy consumed for ice destruction. According to experimental data [55, 65], the application of an electric pulse anti-icing system is efficient for air, spray, and internal water icing on flat, weakly extended surfaces at an ice thickness of more than 4 mm in the case of thin-walled structures. If the ice thickness is less than 4 mm, especially for sea ice, the electric pulse anti-icing method is inefficient. In this case it is necessary to use, for example, thermal anti-icing systems.

There are data on the destruction of ice by puncturing it as a dielectric layer by a high-voltage pulse discharge (1440 kV) [8]. Owing to its low effectiveness, the electric puncture should not be used as a means for the destruction of ice.

4.2. Electric Hydraulic Technique

The main idea of the electrohydraulic effect (EHE) is that a high hydrualic pressure arises around the zone of discharge in the case of a high-voltage pulse spark discharge in a volume of liquid. Mechanical mixing of the liquid in the zone of discharge is accompanied by destruction in other media, because the practically incompressible liquid expands with tremendous force in all directions from the line of discharge and develops a hydraulic impact. The cavity then is compressed at the same speed, developing a second 'cavitation' hydraulic impact. This is a complete cycle, which repeats at a pulse alteration frequency. The destruction of solid bodies occurs as a consequence of the joint action of several factors which develop during high-voltage spark discharge: mechanical impact action of superhigh hydraulic pressure, powerful cavitation processes, luminous radiation, and possible resonance conditions. The destructive effect of the electrohydraulic pulse is similar to the explosive effect of a common explosive charge of equivalent power.

Electrohydraulic plants are designed on the principle of accumulating high-voltage power in a bank of capacitors (up to several dozen KJ) and its subsequent release in the discharge gap of the working member, where electrical energy is transformed into mechanical energy with an efficiency of several dozen percent, for example, when breaking up rock.

The electrohydraulic effect has found widespread application in many branches of machine-building, in mining, construction and agriculture.

Attempts have been made to apply the advantages of this electrophysical method for the generation of mechanical energy and the destruction of ice. The advantages of EHE, in this case, primarily include direct transformation of electric into mechanical power, and the potentiality of simple control of the process of generating energy pulses within a wide range of values and at practically any required frequency.

The first results of the practical application of EHE for the destruction of ice were obtained at the Arctic and Antarctic Research Institute in 1956. The following phenomena were observed during the experiments. On a short explosion pulse in the core of an ice cube of about 1 m^3 the character of ice destruction depended on the medium with which the cube faces had been in contact. If the specimen was surrounded by air, then numerous cracks developed as a result of the explosion and it disaggregated into numerous fragments on the application of a small force. When experiments were carried out on ice specimens of the same dimensions placed in water, the effect of destruction was significantly less: sometimes an explosion resulted in cracks splitting the ice into only two or three fragments. Evidently, there was a substantial reflection of the energy of the stress waves from the ice into the water owing to the good acoustic contact between them. (The experiment was carried out on Lake Ladoga; the energy used for the explosion accumulated in a capacitor and did not exceed a few KJ [17].)

The efficiency of utilization of the accumulated electric power was not more than a few percent in the first experiments on destroying floating ice slabs. It is reported in [161] that attempts were made to increase the efficiency by developing an electrohydraulic monitor with a water jet pressure of 3000 MPa at the outlet. The authors assumed that the efficiency of transforming electric power into hydrodynamic power could reach 90%.

4.3. Laser Technique

The potential use of lasers for ice destruction has been investigated in the USA [4, 201]. A continuous-action CO_2 50 W laser had been used in an experiment on cutting ice speciments made from distilled water and tap water by freezing ice slush particles with a diameter up to 1 cm [201]. The beam was focused by a mirror to a power flux density of 50 W/cm^2. A pulsed laser was also used with a pulse energy of 0.5 J/μs.

The continuous-action laser cut ice with an effectiveness of

3-5 cm^2/s (effectiveness is the area of destruction per unit of time). According to US scientists, a 50-100 W laser can cut an ice field to a depth of 8 cm at a rate of 1852 m/h. The cutting effectiveness drops when the specimen is in a horizontal position, because the meltwater is unable to drain from the cut (the cutting rate drops by a factor of 2, according to our estimates).

The pulsed laser destroyed ice with a maximal thickness of 3 mm. The conclusion had been drawn that it was advantageous to use lasers operating in the near-infrared visible part of the spectrum, focusing radiation not on the surface, but inside the ice, causing a steam explosion therein. The experiments were carried out on artificial ice, but the authors suggested using the laser under field conditions (from an icebreaker) in order to cut or perforate the ice in the zones of maximal stress at the shoulders of the icebreaker to facilitate disintegration of the ice field. It has been found that the laser beam penetrates instantaneously through the snow cover, and the ice-cutting rate is invariable. Explosion-type generation of steam by the energy of the pulse focused under the ice surface considerbly increases the effect of ice sheet destruction. In order to supply the laser energy to the water under the ice with minimal losses in penetrating through the ice sheet, it is necessary to consider the degree of absorption of the beam energy by the ice, i.e. the change in the ice coefficient of absorption when beams with different wavelengths penetrate through the ice.

The laser ray beam gun (pat. 1,119,948 GB) serves the same purpose. The gun concentrates the rays of several lasers into one beam by means of one large concave and one small concave confocal parabolic mirror.

The possibility has been investigated of using a laser for the prevention of ice on structures [225]. Glass lasers (wavelength 1.06 μm) and ruby lasers (wavelength $6943 \cdot 10^{-10}$ m) have been used in the experiment. Stone, asphalt, concrete, brass, aluminium, and steel have been tested as the ice substrate. It has been found that a single pulse with a flux power of 10^8-10^9 W/cm^2 makes radial cracks within a diameter of 0.1 to 2.0 cm in the ice cover. The process of crack formation depends on the focussing as well as on the ice thickness and the content of air bubbles in it, and it is independent of the substrate material.

If it were possible to spread the system of cracks across the iced surface by scanning the beam, the ice could be completely destroyed and a structure totally de-iced.

The use of a laser for melting large volumes of ice seems to be unrealistic because of the heat capacity of the ice.

4.4. Electromagnetic Field, Radioactive Irradiation, Supersonic Waves

Electromagnetic fields of high intensity in the high (h.f.) and super-high (s.h.f.)[*] frequency ranges are widely used in many modern

---

[*] Electromagnetic waves with a frequency of 433.920 MHz and above are in the s.h.f. range; and lower frequencies are in the h.f. range.

technologies in industry, medicine, science, and engineering, when effective volumetric heating of non-metallic materials is required and ordinary contact heating of these materials by means of heat conductivity is ineffective [90, 115].

In the case of interaction of radio-frequency radiation with ice, as with any other dielectric, the basic process in the medium is dielectric heating, and according to the known correlation

$$P_o = 0.555 f \varepsilon' \text{tg} \delta E^2,$$

the specific electromagnetic power P_o (W/cm^3), released in a unit of ice volume at the given intensity of the field E, depends on its frequency f and the electric properties of the material, the loss tangent of the dielectric tgδ and the relative permittivity ε'.

According to L.B. Nekrasov [100, 101], on the exposure of freshwater polycrystalline ice to a strong electromagnetic field (E = 4 kV/cm, f = 40.68 MHz) disturbance occurs in the optical and mechanical homogeneity in 1.5-2 s, inclusions of volumetric water appear in 3-4 s, and intensive cracking of the irradiated specimen commences in 7-8 s, ending in complete decomposition into individual crystals. The absorption of energy occurs mainly on the surface and at the borders of the grain, where there are quasi-liquid film inclusions. The liquid interlayers increase their size up to the appearance of volumetric water, while the mechanical strength decreases.

A powerful s.h.f. electromagnetic field also has a marked effect on the structure and properties of saline sea ice. The ultimate strength of the ice on compression drops considerably at an absorption of power equal to 0.25 W/cm^3, owing to the rapid rise of temperature and intensive draining of the brine from the ice volume resulting in the formation of a porous 'honeycomb' structure, which yields to mechanical force. The effect of the loss of strength by the ice is proportional to its salinity, because an increase of the latter causes a rise of specific absorption of s.h.f. energy in sea ice (Figure 4.1).

Calculation of the effectiveness of high-frequency melting of ice (ε' = 3.45, tgδ = 0.04; initial ice temperature -18°C; volume of specimens ~ 500 cm^3), for example of utilizing a r.f. generator*) with an oscillating power of 600 W at 2.375 GHz, proves that only about 25% of the radiated power was released in the specimen, and the efficiency of the whole system was about 10%. This is associated with the dissipation of electromagnetic waves in a free space, as well as with the fact that a water film forms on the ice surface on melting, having a high value of ε' and thereby 'shielding' the specimen.

It is possible to increase the efficiency, but it should be remembered that it is disadvantageous from the energy viewpoint to melt the entire amount of ice, and especially to heat and evaporate the water. Hence, it is evidently necessary to irradiate the ice with such energy

---

*) This is the power, for example, of the industrially manufactured Straume domestic magnetron equipment with a 1600 W power consumption from the network.

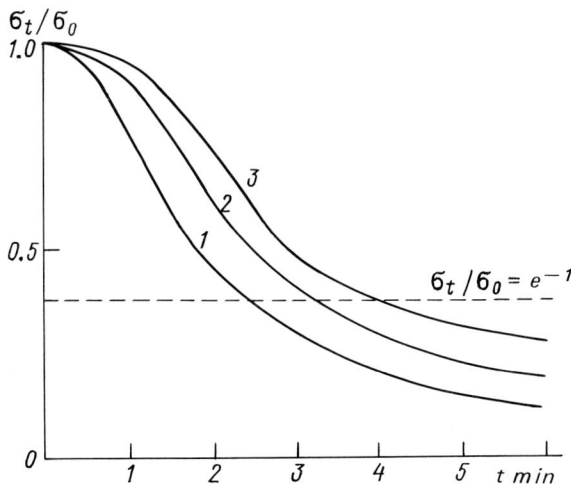

Fig. 4.1. Reduction of relationship of compression strength of an irradiated 444 cm^3 ice specimen (σ_t) to compression strength without irradiation (σ_o), depending on time (data of G.P. Khokhlov and V.V. Pasynkov).

1 - 1°/oo; 2 - 6.8°/oo; 3 - 3.9°/oo.

that the specific absorption of energy in the liquid inclusions should be at least 10^3 W/cm^3, i.e. higher than the values at which practically instantaneous melting of the ice and evaporation of the water occurs.

The calculations indicate that the temperature differential in the spots of local heating may reach 100°C at a frequency range of 10^9-10^{10} Hz. This gradient may cause a heat shock, and a hydraulic impact at the formation of a liquid phase in the local volume, thereby initiating cracking and breaking of the ice at the spots of inhomogeneity.

Preliminary experiments on the positron annihilation technique for studying the effect of irradiation on ice specimens have demonstrated that the effect of the rays in the 4-11 Mrad range is manifested in the migration of voids within the ice at a wide temperature range from -185°C to the melting point [206]. The values of void formation energy have been found to be 0.20-0.35 eV. Their concentration is several ppm [234]. The number of voids forming and the degree of their migration increases with the introduction of hydrogen fluoride. Cavities form in the ice as a result of coalescence of the voids, thereby reducing the strength of the ice [207].

When cutting ice, the application of high-frequency elastic vibration to the cutting tool (ultrasonic generator UZG-2.5 m, 2.5 kW, 18-22 kHz) has made it possible to reduce the effort on the cutter and increase the cutting rate by a factor of 1.5-2 [72].

The degree of reduction of the cutting effort at a fixed density of s.h.f. energy depends on the relationship between the cutting depth and the depth to which the electromagnetic field reaches.

The dynamic loads on the cutting tool reduce by a factor of 2-2.5 when frozen soil is treated with s.h.f. energy [103].

4.5. Inventions for Electrophysical Destruction of Ice

Nos	Country	IPC,NPC	Patent No.	Field	Year of publication	Filed by
1	USSR	E02B,15	549,546	2 Jan. 1973	1977	V.N. Aristov

For ice sludge prevention on hydrotechnical structures by means of ultrasound vibration. To intensify the ice formation process, the area treated by ultrasound vibration is fenced in.

| 2 | USA | E02B,15; B63B,35; 61-103 | 4,077,225 | 28 Feb. 1977 | 1978 | Sun Oil Company |

For ice disaggregation. A system for fragmenting ice round an offshore structure, comprising a transducer for generating acoustical energy, and an elongated beam adapted for engagement with ice. The acoustical energy generated by the transducer is transferred to the beam and is applied to the ice from the beam.

| 3 | Great Britain | E21B; H1C | 1,119,948 | 24 June 1967 | 1968 | Arthur P. Pedrick |

Laser ray beam gun, or concentrator for use in polar regions. A laser ray beam gun or concentrator for the purpose of concentrating rays from several laser devices into a single beam, comprising confocal large and small parabolic mirrors, the rays from the laser devices being directed on to the concave mirror surface of the large parabolic mirror to reflect onto the convex surface of the smaller parabolic mirror, the rays being reflected from the smaller mirror in parallel with its axis in a concentrated beam which passes through an aperture in the large mirror. The smaller mirror is cooled by a fluid supplied and removed by pipes. The apparatus can be mounted on a traversing vehicle and directed downwards to burn a continuous furrow in an ice surface.

Chapter 5

COMBINED DESTRUCTION OF ICE

Often several methods are combined to increase the effectiveness of ice destruction. The ice-breaking system described in pat. 3,698,341, USA, comprises breaking and hydraulic jet methods, and the ice-cutting system in pat. 4,005,666, USA, includes cutting and hydraulic jet methods. The self-powered vehicle in pat. 3,632,172, USA, ingeniously combines cutting, explosive, and electrothermal means for the destruction of ice. This complexity greatly improves the performance of technical means and suggests directions for their subsequent improvement.

Keel vibration of an icebreaker's hull is generated by a combustion chamber in a Canadian technical solution (pat. 998,884 Canada). The lowered bow portion (Figure 5.1), with a plough-like, highly raked configuration goes down below the waterline. A chamber in the bow portion communicates with the external medium through a perforated bottom and valves. A combustion chamber in the adjacent compartment communicates with the former chamber through valves. Compressed air for combustion of

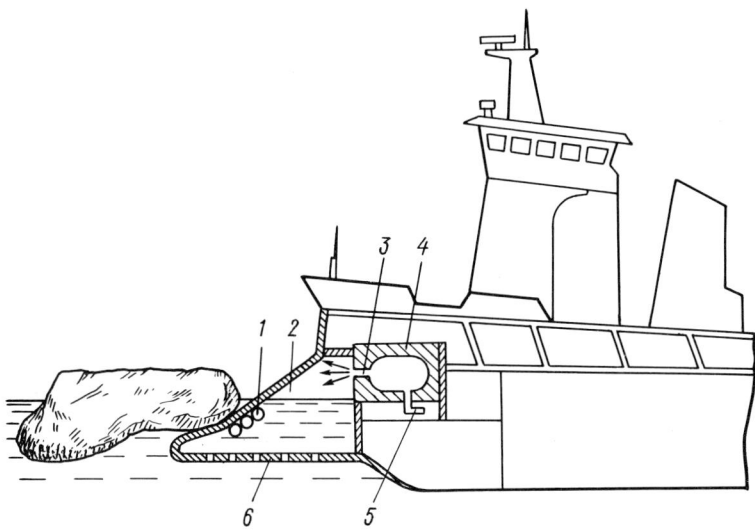

Fig. 5.1. Combined destruction of ice by underwater gas exhausts and keel vibration of the hull [80].

1 - valves; 2 - chamber; 3 - outlet conduit; 4 - combustion chamber; 5 - compressed-air conduit; 6 - perforated bottom.

the fuel is supplied through a conduit. Keel vibration of the ice breaker's hull occurs as a result of cyclic ignition of the fuel in the combustion chamber. Each cycle is accompanied by displacement of the water from the chamber, causing a trim and a reactive force directed upward. The gas discharged into the water additionally weakens the ice sheet, forming cracks and causing sinking of the floes under the effect of their own weight. The chambers are subsequently charged with water through valves. The ice sheet is rapidly broken by the joint action of the vessel's bow portion and the gas pressure.

An icebreaker oil tanker (pat. 3,768,427 USA) is designed for navigating in ice-free and ice-covered waters. The tanker has a wide hull and a narrow superstructure. When navigating in ice, the water ballast should be sufficient for the ice cutters on the ramp to ensure effective destruction of the ice. A jet engine is mounted on the superstructure to increase the navigational capability in ice. A jet of incandescent gas, directed downwards, facilitates breaking of the ice sheet.

The effectiveness of removing the ice cup from the underwater part of the ship's hull is greatly increased by the simultaneous application of several methods. The simplest combined technique is to undermelt the ice cup with steam supplied into the compartments of the ship, accompanied by towing, whereupon the ice cup is washed off by the opposing water current. The removal of the ice cup in this case is approximately twice as fast as when melting it only by supplying steam into the compartments. Nevertheless, the process is time-consuming.

A bottom preheater, equipped with ballast tanks and a thermal member (a steam pipe), may also be referred to as a combined means [50]. The preheater's body is shaped like a bilateral flat wedge with a base sloping in both length and width. When it is necessary to remove the ice cup, the device is arranged at the dockside and the ship is brought close to the thermal member. The ship is connected to a winch by means of a special cable system, the ballast tanks of the preheater are filled up and the ship is arranged over the thermal member (Figure 5.2). When dragging the ship over the thermal member by means of the winch, the ice under the bottom is melted and simultaneously chipped off by the sharpened wedge-like edge of the preheater's body. The chipped-off ice is moved away by a directional current created by a special pump. The device can remove ice only from a flat portion of the ship's bottom, and its manufacture demands substantial capital investments that are justifiable only in the case of numerous dockings of flat-bottomed ships affected by ice cups.

Flow preheaters can be used for removing the ice cup [27]. Experiments on removing the ice cup were carried out with steam flow preheaters on the Severnaya Dvina River in Arkhangelsk in 1971. The flow preheaters were fixed on rafts in front of the bow portion of a tanker which had an ice belt at the waterline and a powerful ice cup on the underwater part of the hull. However, the experiment was not completed owing to lack of time and consumption of steam. The ice cup and the ice belt were only one-third melted along the ship's length in its bow portion in 42 hours.

Chipping and electrothermal means have been combined to develop a

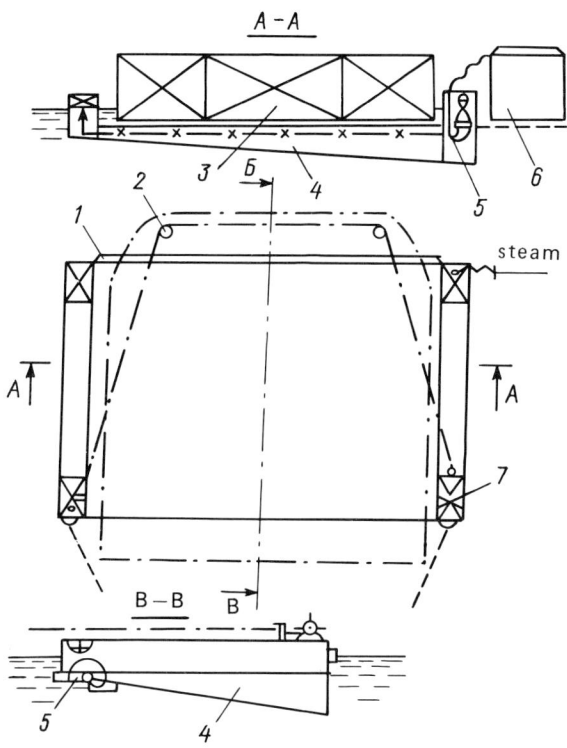

Fig. 5.2. Combined device for removing ice cup from a ship's bottom [50].

1 - thermal member; 2 - cable system; 3 - ship; 4 - preheater's bottom; 5 - pump with nozzle; 6 - pier; 7 - winch.

balance scraper. The assembly comprises two symmetrical members (scrapers), forming an angle in the plane with the apex in the direction of navigation and coinciding with the diametral plane of the dock. Each element of the scraper is mounted on a live axle, which rests on the bearings of the central and lateral foundations. The foundations are equipped with hoisting mechanisms, to allow for the deadrise of the ship's hull. The bottom part of the scrapers comprises a distributed load, serving as a counterweight and ensuring that the scraper's upper blade fits the ship's bottom even with a change of draught along the ship's length owing to trim. The scraper's blade comprises a replaceable wooden bar-shock absorber and a conduit with nozzles to supply water, steam, or compressed air. The scrapers are detachable for convenience of operation, and the foundation is of a knock-down design. The floes removed from the bottom by the scrapers are moved to the sides of the ship and surface.

A combined system proposed in Norway comprises three individual devices, the use of which depends on the parts of the ship protected against icing. The masts, booms, yards, and other hollow masting are equipped with tubular heating elements, comprising infrared heat sources. The solid masting (forged and rolled), and the rigging and antennas are protected by external heating elements.

The handrails, manropes, stanchions, davits, and part of the rigging and deck mechanisms are coated with a special low-adhesion polymer (phortiflex). The bridge and superstructure are protected by an electrothermal anti-icing system [1].

The anti-icing devices should have mainly vertical surfaces. It is advisable to apply the anti-icing coatings on surfaces which are not protected by electric heating covers, to prevent the accumulation of ice on the forecastle and not to bring down the bow: the forecastle, sides, and bulwark up to the 50th freame, the side bulwark, superstructures, trawl gallows.

An ice-cutter (pat. 2,341,932 FRG) is assembled on icebreakers to cut a trapezoidal trench ahead of the ship's stem. This is in the form of sleds with three planing cutters of different length and width. When the ship moves ahead, the cutters cut off parallel slices of ice, thereby making a trench in the ice field. The ice cutters are suspended from a bearing column by means of hinged tie-rods. A vibrator fixed on the column generates high-frequency vibration in the cutters, greatly increasing their performance. The cutters can be heated to prevent the cutter freezing up in the ice.

The so-called electric thermomechanical method of working frozen rock is currently becoming widespread in practice [34, 99]. While high-frequency dielectric heating is used in traditional areas as an independent energy factor, as described above, high-frequency heating is also important in intensifying the development of frozen rock and ice. As such, it is organically associated with another technological stage in the destruction process, affecting material that has already lost its strength.

This combined method was used for the first time in an electric thermomechanical system developed by L.B. Nekrasov, used in sinking a slant shaft in artificially frozen rock in the construction of a Metro station [103].

The fact that ice can be effectively destroyed in electromagnetic fields has been exploited in another means for ice drilling (Author's Certificate 247,187 USSR).

Combined methods of ice destruction include the method of making lanes by sawing out ice blocks with a mechanical or thermal cutter and subsequently sliding them under the ice cover. The method can be used for preparing unloading areas on fast ice, when chipping ice off ships, in various underwater operations, fishing under ice, etc.

The essence of the method is as follows. One or two ice blocks are cut out of the ice; one of the cuts is skewed, and the ice block is then slid under the ice cover by means of some hauling device. When two ice blocks are cut out, they are slid one under the other. Hence, a lane in the ice and an ice area of double thickness are made simultaneously. The effort required for sliding the block under the ice, or for sliding one block under the other, considering the coefficient of friction of ice agianst ice to be 0.1, is approximately 10 times less than the effort required for hauling the block out of the ice surface, and 40-50 times less than the effort required for direct extraction of the block (by means of some hoisting mechanism). With a sufficient depth of the water under the ice and minor snow cover it is more advisable to push the block under the ice cover, thereby reducing the number of cuts. When the snow cover is deep and the depth of the water under the ice is small, it is more advisable to cut out two similar ice blocks and push one on top of the other. The sinking of the ice blocks will be minimal, rendering it possible to reinforce the bearing capacity of the ice cover directly at the shore on shallow waters for the construction of unloading areas.

Ice blocks can be cut out mechanically, for example, by means of an ice milling cutter machine, comprising a system of milling cutters on a cross-country vehicle, or by a hydraulic thermodrilling unit, which cuts the ice by a row of parallel water or steam jets.

5.1. Inventions for Combined Destruction of Ice

Nos	Country	IPC,NPC	Patent No.	Filed	Year of publication	Filed by
1	USSR	65a^1,8	217,979	16 Jan. 1967	1968	I.L. Rabey

For removing an ice cup under a ship's bottom. Comprises a bottom preheater with ballast tanks and a tubular thermal member. To simplify the design and reduce heat consumption, the preheater's body is in the form of a bilateral flat wedge with its bottom sloping in both width and length, and the thermal member is arranged by the edge of the extremity in the sharpened part of the wedge. Together they have an effect on the ice cup when the ship is dragged over the preheater's deck.

| 2 | USSR | B63B,35 | 481,495 | 14 Mar. 1972 | 1975 | I.V. Zagryadsky, B.Ya. Myalkin, Yu.D. Kravchuk |

Ice-cutting means. The working member is a hollow rod with conical rings, having cutting edges. The rod comprises openings for the passage of steam.

| 3 | USSR | E21c,6 | 554,401 | 3 June 1971 | 1977 | E.A. Ignatenko |

Method of rotary drilling. The tool is rotated, axial force is imparted, and a washing fluid is supplied simultaneously under pressure to the cutting zone. The washing fluid is imparted with ultrasound vibrations to raise the effectiveness of drilling.

| 4 | USSR | E02d,5; B63B,45 | 675,132 | 19 Dec. 1977 | 1979 | V.N. Pikul |

Impact means with a combustion chamber for ice destruction. Comprises a working cylinder, fuel system, means for starting, and working member.

The latter member is in the form of a sharpened cutting edge on the bottom face of the working cylinder to raise the effectiveness.

| 5 | USA | B63B,35; 114-40 | 3,572,273 | 6 Aug. 1969 | 1971 | Southwest Research Institute |

Apparatus for breaking a layer of ice on a body of water by repetitive combustive explosions. The apparatus comprises a movable buoyant body having a face for contact with the ice and several exhaust openings in the face and positioned below the waterline of the body with each exhaust opening connected to a combustion chamber for applying the combustion energy from a hydrocarbon fuel directly to the ice for breaking and melting the ice.

| 6 | USA | B63B,35; 299-13 | 3,632,172 | 17 July 1969 | 1972 | Dresser Industries Inc. |

For weakening ice to assist an icebreaker. A self-powered vehicle for moving across the ice ahead of an icebreaker and carrying means for cutting a trench along the path of the icebreaker and for placing an explosive cord in the trench which can be exploded for weakening the ice. An air cushion vehicle carries a rotary saw and means for removing ice cuttings from the trench made by the saw, with provision for melting part of the ice cuttings for backfilling along the trench after inserting an explosive cord. Equipment is carried by the air cushion vehicle for drilling spaced holes along the ice for additional explosive charges.

| 7 | USA | B63B,35; 114-40 | 3,698,341 | 4 Mar. 1970 | 1972 | Jacob C. Wagner |

Icebreaking system for ships. Comprises a compressed-air operated ram which is movably attached to the ship. The ram moves freely beneath the water. The ram comprises three compartments. The uppermost compartment provides for breaking the ice from beneath, and is connected with a compressed air means to throw the ice away by expelling water under high pressure. The intermediate compartment is connected with the compressed air means to drive the ram by rapidly expelling sea water downwardly from conduits. The lowermost compartment, like that of a diving bell, is a buoyancy chamber containing some volume of air.

| 8 | USA | B63B, 35/12; 114-41 | 3,768,427 | 30 Oct. 1970 | 1973 | Robert Melling Stephens |

Icebreaker oil tankers. A tanker characterized by three large cast steel saws disposed forward of the narrow superstructure which undercut the ice. Since the vessel is designed to navigate with the assistance of a conventional icebreaker, supplementary water ballast compartments additional to the regular ballast tanks are incorporated on the bottom of the hull.

| 9 | USA | B63B,35; 114-40 | 3,934,529 | 30 Oct. 1974 | 1976 | Canadian Marine Drilling Ltd. |

Icebreaking vessels. A procedure for breaking ice which comprises the following cyclic steps: pretensioning an ice sheet by plough under a portion of the ice sheet with a plough portion of the bow of an icebreaking vessel; providing a mass of water within a chamber disposed within the plow portion of the bow; applying exhaust gases from a combustion chamber to the mass of water in the chamber. The gases have a pressure of at least 700 psi, thereby rapidly expelling the water. The downwardly expelled water develops an upward reactive force superimposed on the pretensioned ice sheet.

| 10 | USA | B63B,35; 114-42 | 4,005,666 | 23 June 1975 | 1977 | Sea-Log Corporation |

Fluid vacuum release for ice-cutting systems. A comminuting ice-cutter comprising several rotating cutter elements, the cutter elements having cutter edges spaced circumferentially around a common axis and adapted to engage ice and dislodge pieces of ice by impacting action; a drive rotating the cutter elements in a direction around the common axis to apply high velocity impacts to the adjacent ice; and means directing fluid under pressure at the interface between each of the cutting edges of the cutter elements and the ice when it comes in contact with the cutting edges, the fluid means dissipating the partial vacuum resulting from ice cleavage by fluid ejection and changing it into a positive pressure which aids the separation and removal of the chips by the cutter elements.

| 11 | USA | B63B,35; 114-40 | 4,083,317 | 11 Feb. 1977 | 1978 | John E. Holder |

A water discharge system for breaking ice, comprising an inlet conduit positioned to remove the water from beneath the ice, and an outlet conduit having laterally and downwardly extending portions both of which provide a discharge opening for discharging water onto the surface of the ice at a volumetric rate sufficiently great to provide a load heavy enough to break the ice by the effect of the weight of the accumulated water and the thermal stress created within the ice by the water.

| 12 | Canada | 114-11 | 998,884 | 16 Sept. 1974 | 1976 | Canadian Marine Drilling Ltd. |

Icebreaker. A vessel comprising a hull having a bow portion of plough-like, highly raked configuration, a single water chamber disposed within the plough-like bow portion and communicable through the hull below the waterline. A combustion chamber in the adjacent compartment. Compressed air is supplied through a conduit. High-velocity gas from the combustion chamber communicates directly with the water chamber and cyclically displaces water contained therein rapidly downwards. Each cycle is accompanied by displacement of the water from the chamber, causing a trim and a reactive force directed upwardly. The ice sheet is rapidly broken by the joint action of the vessel's bow portion and the gas pressure.

REFERENCES

1. L.R. Aksyutin. Icing of Ships. Leningrad, Sudostroyenije Publishers, 1979.
2. D.D. Alimov, I.G. Basov, F.F. Zelinger. Experience of using a jib machine for ice cutting. Stroitelniye i Dorozhniye Mashiny, 1969, No. 2, pp.11.
3. B.A. Amelin. Technology of hole drilling in the Arctic and Antarctica abroad. Kolyma, 1976, No. 4, pp.31-34.
4. M. Amusin, N. Obolenskaya. Internal waterways in the USA. Rechnoy Transport, 1978, No. 11, p.47-51.
5. N. Antrushin. Radiation-chemical method of disintegrating spring ice. Proceedings of Leningrad Institute of Water Transport Engineers, 1962, Issue 30, pp.15-17.
6. N. Antrushin. Aviation-chemical methods of accelerating ice melting. Proceedings of Coordination Conferences on Hydraulic Engineering. Moscow, 1965, Issue 17, pp.219-223.
7. A.V. Arbatsky, V.G. Vionich, R.G. Solomin. Hand vibro-impact tools for freezing-out operations. Proceedings of Research Institute of Water Transport, Issue 126, pp.57-62.
8. A.V. Astafurov. Electric puncture of thick ice layers by pulses. Proceedings of USSR Academy of Sciences, Tekhnicheskaya fizika, 1953, Vol. 22, No. 4, pp.419-422.
9. A.V. Bazhev, Regularities of melting an artificially dusted glacier surface. Methods of Glaciological Research. News items, discussion, 1973, Issue 21, pp.124-125.
10. V.V. Balanin. Maintenance of prolonged operation of slips. Proceedings of Leningrad Research Institute of Water Transport, 1965, Issue 83, pp.7-18.
11. V.V. Balanin, B.S. Borodkin, M.I. Zhidkikh. Field tests of experimental plant to maintain prolonged slip operation. Proceedings of Leningrad Research Institute of Water Transport, 1965, Issue 83, pp.65-72.
12. V.V. Balanin, B.S. Borodkin, G.I. Melkonyan. Utilization of Heat of Deep Waters in Water Bodies. Moscow. Transport Publishers, 1964.
13. V.N. Bartenyev. Ice cutting machine for heavy ice. Proceedings of Research Institute of Water Transport, 1977, Issue 126, pp.44-49.
14. B.D. Bass. Condition and potentiality of improving equipment for making lanes in ice at operation of suction-tube dredges in early spring. Proceedings of Research Institute of Railway Engineers, 1978, Issue 135, pp.67-84.
15. E.V. Byelysheva. Selection of chemical substance causing ice melting. Hydrometeorological Conditions of Ship Icing. Leningrad, Gidrometeoizdat Publishers, 1969.

REFERENCES

16. V.V. Bogorodsky, V.P. Gavrilo, A.V. Gusyev. On nonlinear effects at ice destruction in liquid. Proceedings of Arctic and Antarctic Research Institute, 1970, Vol. 295, pp.159-165.
17. V.V. Bogorodsky, V.P. Gavrilo, P.V. Ponomaryev. Dependence of strength of solid bodies on acoustic contact with boundary media. Proceedings of Arctic and Antarctic Research Institute, 1974, Vol. 324, pp.80-90.
18. V.V. Bogorodsky, V.P. Gavrilo, V.I. Ivanov. Investigation of plastic deformation of freshwater ice at an impact load. Proceedings of Arctic and Antarctic Research Institute, 1980, Vol. 374, pp.43-50.
19. V.V. Bogorodsky, V.P. Gavrilo. Ice. Physical Properties. Modern Methods of Glaciology. Leningrad. Gidrometeoizdat Publishers, 1981.
20. B.S. Borodkin. On calculation of perforated air pipelines. Proceedings of Leningrad Institute of Water Transport Engineers, Issue 24, pp.62-66.
21. B.S. Borodkin, M.I. Zhidkikh. Methods of fighting against ice on water areas of some slips of Ministry of River Fleet. Proceedings of Leningrad Institute of Water Transport Engineers, 1967, Issue 88, pp.16-20.
22. Prevention of ice on roofings. Express Information of Central Bureau of Scientific-Technical Information of Leningrad Communal Economy of RSFSR, Housing Economy Series, 1976, No. 28.
23. Yu.I. Bublikova. On use of radiation method for weakening ice effect on hydrotechnical structures. Izvestiya Vuzov, Construction and Architecture Series, 1971, No. 4, pp.133-138.
24. Yu.I. Bublikova, A.F. Kamynina, M.V. Kunyavskaya. Intensity of melting ice cover under natural conditions and at blackening it. Proceedings of Novosibirsk Regional Hydrometeorological Centre, 1969, Issue 2, pp.138-143.
25. V.E. Buchinsky. Ice Sleet and its Prevention. Leningrad, Gidrometeoizdat Publishers, 1960.
26. N.F. Buyanov. How to fight against icing of ships? Rybnoye Khozaistvo, 1967, No. 2, pp.30-37.
27. L. Bykov, V. Klyuyev. Ship locks on Saint Lawrence Seaway. Rechnoy Transport, 1976, No. 2, pp-53-55.
28. L.M. Byalobzhesky, L.M. Rudakov. Modern methods of fighting against slipperiness on roads. Avtomobilniye Dorogy, 1972, No. 9, pp-12-14.
29. L.V. Valkovsky, S.I. Levin, A.I. Sezin. Use of double-jig machine in ice cutting and blasting operations. Stroitelstvo Truboprovodov, 1970, No. 7, pp.28-29.
30. S.A. Vershinin. Break and load-carrying capacity of ice field at brief static and quasi-static loads. Ice Thermal Phenomena and Their Consideration in Building and Operating Integrated Water Power Developments and Hydrotechnical Structures. Leningrad, 1979, pp-92-96.
31. E.S. Vinogradov. Equipment of dry docks with anti-icing means. Sudoremont Flota Rybnoy Promyshlennosty, 1972, No. 20, pp.36-41.
32. V.G. Voinich. Ice destruction by hydro-impact tools. Proceedings of Research Institute of Water Transport, 1977, Issue 126, pp.50-56.
33. Vologda ice drill. Rybovodstvo i rybolovstvo, 1964, No. 4, pp.38-39.
34. L.V. Nekrasov, Yu.Yu. Misnik, V. Dobretsov, V. Fateyev, High-frequency knife for frozen soil. Izobretatel i Ratsionalizator, 1971,

No. 2, pp.20-21.
35. V.P. Gavrilo, A.V. Gusev, Yu.K. Zaretsky, A.M. Fish, Acoustic emission as index of ice deformation and destruction. Proceedings of Arctic and Antarctic Research Institute, 1978, Vol. 359, pp.118-126.
36. A.I. Glebov. Ice destruction by cutting in medium of surfactants. Proceedings of Research Institute of Water Transport, 1977, Issue 126, pp.34-44.
37. A.I. Glebov, N.N. Monzyrev. Some experimental results of ice destruction by cutting. Proceedings of Research Institute of Water Transport, 1978, Issue 135, pp-24-31.
38. S.G. Gomolsky, N.G. Khrapaty, V.G. Tsuprick. Investigations of impact of solid body on ice. Ice Thermal Phenomena and their Consideration in Building and Operating Integrated Water Power Developments and Hydrotechnical Structures. Leningrad. 1979, pp-73-77.
39. V.F. Grishchenkov. Fighting ice slush on Moscow—Leningrad highway. Avtomobilniye Dorogy, 1973, No. 9, pp.14-15.
40. A.A. Gundobin. Fighting Icing of Ships. Vladivostok. Far East Book Publishing House, 1966.
41. A. Gundobin, M. Kuznetsov. Fighting icing of tankers. Morskoy Flot, 1972, No. 11, pp.27-28.
42. A.D. Dolgushin, I.A. Zotikov, E.N. Tsykin. Problem of artificial regulation of ice and snow melting. Engineering Glaciology. Apatites, 1973, pp.218-222.
43. L.D. Dolgushin, G.B. Osipova, O.V. Rototayeva. Experience of artificially enhancing ice and snow melting by radiation method. Materials of Glaciological Investigations, 1976, Issue 27, pp.187-195.
44. A.M. Yestifeyev, A.N. Pekhovich, S.M. Aleinikov. Blackening surfaces of ice cover as a method for accelerating ice melting in spring. Proceedings of Research Institute of Hydraulic Engineering, 1960, Vol. 65, pp.139-148.
45. O.N. Zhabotinsky. Use of cargo and specialized hovercraft abroad. Advanced Experience and New Technology, 1978, No. 4, pp.67-72.
46. I.V. Zagryadsky, Yu.D. Kravchuk. Methods for destruction of ice cover at creation and maintenance of lanes in freezing sea ports. Proceedings of Coordination Conferences on Hydraulic Engineering (additional materials), 1973, Issue 81, pp.62-64.
47. Protection of internal waterways against freezing by water heating. Reference Journal Water Transport, 1965, No. 10B730.
48. Yu.I. Zvaizne. Light ice drilling unit. Rybnoye Khozaistvo, 1966, No. 6, pp.41-43.
49. F. Zelinger. Jig machine for cutting ice. Rechnoy Transport, 1981, No. 1, p.14.
50. L.V. Ivanov, Winter Operation of Water Transport. Moscow, Transport Publishers, 1978.
51. L.V. Ivanov, Artificial heating of water bodies. Sudostroyeniye, 1969, No. 5, pp.55-59.
52. L.V. Ivanov, A.T. Lupa. Operation of mooring structures in ports under icing conditions. Central Bureau of Scientific-Technical Information of Ministry of Merchant Navigation. Review Information, Sea ports, Series; 1975.

REFERENCES

53. Z.M. Ilyin. Ice Drill of P.I. Pshenichnikov and I.N. Morozov for Mechanization of Fishing under Ice. Moscow, Ministry of Fishing Industry, 1956.
54. Instructions for Fighting Ice Slush on Automobile Roads. Moscow, Transport Publishers, 1975.
55. Use of pulse methods in anti-icing devices on vessels of fishing fleet. Collection of Summaries of Research Works at Moscow Technical Institute of Marine Instrument Engineering, 1979, Series 15, No. 5.
56. Yu.P. Kazankov, V.P. Khokhlov. Experience of blackening ice on road testing ground of trust Zabaikalzoloto. Kolyma, 1970, No. 3. p.12.
57. M.O. Kamyshev. Ice picks and ice drills. Rybolov-Sportsman, 1958, No. 10, pp.240-244.
58. G.L. Karaban, V.B. Ratnikov. New technology for mechanized removal of ice and snow from roads. Avtomobilniye Dorogy, 1972, No. 12, pp.122-123.
59. V.N. Karnovich, V.I. Sinotin, Some ideas on fighting ice jams by bombing and blasting operations. Proceedings of Coordination Conferences on Hydraulic Engineering, 1970, Issue 56, pp.65-69.
60. V.N. Karnovich. Jams on Dnestr River in winter of 1966/67 and 1968/69 and measures for their prevention. Proceedings of State Hydrological Institute, 1971, Issue 187, pp.132-144.
61. V.I. Kashtelyan. Pseudo-washing device - a means to increase navigational capability of ships in ice. Sudostroyeniye za rubezhom, 1978, No. 8, pp.55-88.
62. V.I. Kashtelyan, T.X. Yarovaya. Use of means on air cushion for ice destruction. Sudostroyeniye za rubezhom, 1978, No. 5, pp.57-64.
63. R.N. Bobrov, M.P. Latko, V.K. Savinykh, L.Ya. Churakov. On problem of mechanizing freezing-out operations. Proceedings of Research Institute of Water Transport, 1976, Issue 119, pp.40-44.
64. V.I. Kovalenko, V.S. Moiseyev, E.A. Zagrivny. Drilling - melting holes at Vostok-1 station. Proceedings of Soviet Antarctic Expedition, 1981, Vol. 74, pp.112-116.
65. R.T. Kozlovskaya, A.V. Panyushkin. Investigation of effectiveness of pulse destruction of ice. Proceedings of Coordination Conferences on Hydraulic Engineering, 1976, Issue 11, pp.170-175.
66. A.P. Kolosov, A.V. Panyushkin. Change of strength properties of freshwater ice by physico-chemical method. Proceedings of Coordination Conferences on Hydraulic Engineering, 1976, Issue 111, pp.199-204.
67. G. Colsky. Stress Waves in Solid Bodies. (Translated from English.) Moscow, Foreign Literature Publishing House, 1955.
68. V.A. Konovalov, V.M. Falchevsky. Wider use of chlorides to fight ice sleet. Avtomobilniye Dorogy, 1973, No. 9, pp.14-15.
69. I.M. Konovalov. Electrothermal Ice Cutting. Moscow, Narkomrechflot Publishers, 1946.
70. I.M. Konovalov, K.S. Yemelyanov, P.N. Orlov. Fundamentals of Ice Engineering of River Transport. Leningrad-Moscow, USSR Ministry of River Fleet, 1952.
71. E.S. Korotkevich, L.M. Savatyugin, V.A. Morev. Through drilling of shelf glacier at Novolazarevskaya station. Information Bulletin of Soviet Antarctic Expedition, 1978, No. 98, pp.49-52.

72. V.P. Kostylev, L.S. Dyachuk. Investigation of ultrasound cutting of frozen soils. Stroitelniye i Dorozhniye Mashiny, 1972, No. 1, pp.33-35.
73. V.M. Kotlyakov, D.M. Ushakov, V.T. Khodakov. Modern problems of engineering glaciology and economic activity of people. Materials of glaciological Investigations, 1981, Issue 40, pp.211-223.
74. L.N. Krasilnikova, N.P. Kharitonov, V.S. Ivanov. Anti-icing coatings of organo-silicate materials. Heat-Resistant coatings. Leningrad, Nauka Publishers, 1969, pp.379-381.
75. A.N. Krenke. Role of investigating glaciers in studying interaction of man with environment. Investigation and Protection of Hydrosphere. Moscow, 1975, pp.20-23.
76. B.B. Kudryashov, V.F. Fisenko. Analysis and ways of improving drilling-melting process in Antarctic ice. Proceedings of Soviet Antarctic Expedition, 1972, Vol. 60, pp.129-143.
77. V.S. Kulikov. Specific features of blasting ice and frozen rock. Kolyma, 1971, No. 2, pp.15-36.
78. V.V. Lavrov. Deformation and Strength of Ice. Leningrad, Gidrometeoizdat Publishers, 1969.
79. I. Levin. Anti-icing pulse. Izobretatel i Ratsionalizator, 1971, No. 2, pp.7-8.
80. V.M. Levitsky. Means for ice destruction. Advanced Experience and New Technology, 1979, No. 4 (64), pp.57-69.
81. Hovercraft- an ice speciality. Morskoy Flot, 1976, No. 2, p.62.
82. A.V. Abramov, V.G. Aleksandrov, V.V. Manukyan, Yu.I. Strezhnev. Icebreaker-building in patents. Sudostroyeniye, 1976, No. 2, p.38-40.
83. D.L. Lezin. Methods of cleaning navigation canals of ice. Proceedings of Research Institute of Water Transport, 1976, Issue 119, pp.45-92.
84. D.L. Lezin. Icebreaking attachments. Proceedings of Research Institute of Water Transport, 1977, Issue 126, pp.3-15.
85. D.L. Lezin, Yu.D. Namestnikov. Analytical review of icebreakers. Proceedings of Research Institute of Water Transport, 1978, Issue 135, pp.3-12.
86. V.D. Likhomanov, D.E. Kheisin. Experimental investigation of impact of solid body on ice. Problems of Arctic and Antarctic, 1971, Issue 38, pp.105-111.
87. I.M. Konovalov, A.I. Chekrenev, V.V. Balanin, B.S. Antonov. Methods for prolonging navigational period on internal waterways. Proceedings of Leningrad Institute of Water Transport, 1963, Issue 46, pp.30-37.
88. Methodological Instructions for Fighting Ice Jams and Blocking. Leningrad, Energiya Publishers, 1970.
89. E.V. Milovanov, L.G. Tsoy. Prospects of using amphibious ships on air cushion abroad. Sudostroyeniye, 1976, No. 4, pp.67-72.
90. Yu.M. Misnik, L.B. Nyekrasov. Electromechanical destruction of frozen rock. Physico-technical Problems of Developing Useful Minerals, 1973, No. 5, pp.27-30.
91. A. Mogutnyev. Infrared scraper for gas-air sweep. Izobretatel i Ratsionalizator, 1970, No. 12.

REFERENCES

92. P. Molen. Torpedo with thermal head for destruction of icebergs, their fragments, and various masses of ice. Bulletin of Foreign Scientific-Technical Information of Arctic and Antarctic Research Institute, 1971, No. 21 (118), pp.13-14.
93. V.A. Morev. Experiments on drilling ice by electrothermal method at Mirniy. Information Bulletin of Soviet Antarctic Expedition, 1966, No. 56, pp.52-56.
94. V.A. Morev. Effectiveness and economy of electric thermodrilling projectiles on drilling continental ice sheet. Proceedings of Arctic and Antarctic Research Institute, 1972, Vol. 255, pp.158-165.
95. V.A. Morev. Electric thermodrills for drilling holes in ice cover. Materials of Glaciological Investigations, 1976, Issue 28, p.118-120.
96. V.A. Morev, V.A. Shamontyev. Experimental drilling of glacier cover. Information Bulletin of Soviet Antarctic Expedition, 1970, No. 78, pp.102-104.
97. A.A. Nazarovsky. Effect of operating conditions and some design parameters of end milling cutter on process of ice milling. Proceedings of Gorky Polytechnical Institute, vol. 27, Issue 8, pp.21-25.
98. V.D. Namestnikova, N.P. Monzyrev, L.Ya. Churakov. Mechanization of ice chipping operations on ships and trains. Proceedings of Research Institute of Water Transport, 1976, Issue 119, pp.35-39.
99. L.B. Nyekrasov. High-frequency electric thermomechanical devices. Exhibit at USSR Exhibition of National Economic Achievements, Leningrad, 1968.
100. L.B. Nyekrasov. Ice behaviour in quick-changing electromagnetic fields of high intensity. Proceedings of Arctic and Antarctic Research Institute, 1975, Vol. 326, p.90-93.
101. L.B. Nyekrasov, V.G. Ivanov. Ice destruction in strong HF electromagnetic fields. Physico-Technical Problems of Developing Useful Minerals, 1975, No. 5, pp.26-28.
102. L.B. Nyekrasov. Fundamentals of Electric Thermomechanical Destruction of Rock. Novosibirsk, Nauka Publishers, 1979.
103. L.B. Nyekrasov, V.A. Kholinsky. Intensification of frozen rock cutting operations by SHF energy. Kolyma, 1976, No. 1, pp.6-8.
104. A.A. Nikolayev. Investigation of mechanisms of freezing-out operations. Proceedings of Gorky Polytechnical Institute, 1975, Vol. 31, Issue 8, pp.26-28.
105. A.F. Nikolayev. Investigations and Complex of Machines for Developing Frozen Soils, Ice, and Snow. Gorky, 1962.
106. A.F. Nikolayev, Yu.B. Galkin, A.P. Kulyashov. Testing new ice cutting machine. Rechnoy Transport, 1970, No. 10, p.48.
107. A.F. Nikolayev, A.P. Kulyashov. Rotor-Screw Amphibious Vehicles. Gorky, Volga-Vyatskoye Book Publishers, 1973.
108. A.F. Nikolayev, A.A. Nazarovsky. Buoyant ice cutting machines. USSR Scientific-Technical Society, 1969, No. 7.
109. A.F. Nikolayev, V.N. Khudyakov. Self-propelled ice cutting unit. Rechnoy Transport, 1974, No. 3, p.51.
110. S.E. Nikolayev. Experience of blasting shore fast ice at Mirniy. Information Bulletin of Soviet Antarctic Expedition, 1964, Issue

46, pp.37-40.
111. S.E. Nikolayev. Blasting shore fast ice in Antarctica. Information Bulletin of Soviet Antarctic Expedition, 1971, Issue 51, pp.84-88.
112. S.E. Nikolayev. Levelling surface of perennial ice by hole blasting. Problems of Arctic and Antarctic, 1970, Issue 35, pp.105-108.
113. S.E. Nikolayev. Experience of sea ice blasting by a directional explosion. Proceedings of Arctic and Antarctic Research Institute, 1971, Vol. 300, pp.177-195.
114. A.F. Nikolayev, A.O. Vaganov, Yu.B. Galkin, A.P. Kulyashov. New ice cutting machines. Rechnoy Transport, 1974, No. 12, pp.50-51.
115. New Physical Methods for Destruction of Mineral Media. Moscow, Nedra Publishers, 1970.
116. M. Nozdrin. Again on ice drill Ring. "Rybovodstvo i Rybolovstvo, 1969, No. 6.
117. E.P. Borisenkov, M.A. Kuznetsov, G.A. Zablotsky, V.V. Panov. Potentialities of using thermal methods to prevent icing of ships. Proceedings of Arctic and Antarctic Research Institute, 1975, Vol. 317, pp.92-98.
118. V.A. Korenkov, G.A. Morozov, A.F. Nikolayev, A.I. Shkoda. Experience of using ice cutting milling machines for prevention of ice formation. Gidrotekhnicheskoye Stroitelstvo, 1975, No. 2, pp.42-45.
119. E.P. Borisenkov, V.V. Panov, A.V. Panyushkin, Z.I. Shvaistein. Means for protection of ships against icing. Proceedings of Arctic and Antarctic Research Institute, Vol. 317, pp.4-12.
120. V.V. Panov. Icing of ships. Proceedings of Arctic and Antarctic Research Institute, 1976, Vol. 334.
121. V.V. Panov, A.V. Panyushkin, Z.I. Shvaistein. Experimental investigation of ice adhesion under laboratory and field conditions. Proceedings of Arctic and Antarctic Research Institute, 1975, Vol. 326, pp.147-154.
122. E. Pounder. Ice Physics. Moscow, Progress Publishers, 1967.
123. A.A. Pashchenko. Organo-silicon coatings of cold hardening. Kiev, Vyshcha Shkola Publishers, 1972.
124. I.S. Peschansky. Works of ice laboratory of Arctic and Antarctic Research Institute on problem of ice cover destruction. Proceedings of Coordination COnferences on Hydraulic Engineering, 1965, Issue 23, pp.83-84.
125. I.S. Peschansky. Ice Science and Ice Engineering. Leningrad, Gidrometeoizdat Publishers, 1967.
126. V.S. Pelevin. Method of drilling holes in ice by means of high-temperature gas jet. Information Bulletin of Soviet Antarctic Expedition, 1964, No. 48, pp.35-37.
127. E.V. Petrakov. Trends of developing hovercraft for arctic regions. Sudostroyeniye za Rubezhom, 1973, No. 3 (75), pp.24-30.
128. I.G. Petrov. Use of deep water heat to create nonfreezing water areas. Proceedings of Arctic and Antarctic Research Institute, 1964, Vol. 267, pp.81-88.
129. I.G. Petrov. Use of steam to make lanes in ice cover. Proceedings of Arctic and Antarctic Research Institute, 1964, Vol. 267, pp.100-104.
130. I.G. Petrov. Reduction of ice strength by absorption of ambient

REFERENCES

water. Proceedings of Arctic and Antarctic Research Institute, 1976, Vol. 331, pp.50-56.
131. N. Popov. Ice drill Ring. Rybovodstvo i Rybolovstvo, 1966, No. 6, pp.28-39.
132. S.V. Proskurnikov. Formation of artificial canals and lanes in ice. Moscow-Leningrad, Goslesbumizdat Publishers, 1959.
133. B.M. Proskuryakov. Investigation of thermal regime of water bodies and water flows. Proceedings of Conference on Modern Methods of Calculating and Simulating Temperature Fields of Water Bodies. Leningrad, Gidrometeoizdat Publishers, 1966, pp.4-9.
134. A.V. Panyushkin, Z.I. Shvaistein, N.A. Sergacheva, V.B. Rosenzveig, A.P. Petrov, Yu.B. Petrov. Results of testing some means for prevention of icing under field conditions. Proceedings of Arctic and Antarctic Research Institute, 1972, Vol. 298, pp.91-96.
135. L.M. Rudakov. Fighting ice slush on roads by means of brine. Avtomobilniye Dorogy, 1972, No. 9, pp.12-14.
136. E.P. Semyonova. Laboratory testing of chemical reagents of anti-icing means in cold room of Arctic and Antarctic Research Institute. Proceedings of Arctic and Antarctic Research Institute, 1969, Vol. 298, pp.97-164.
137. Yu.A. Simonov, L.B. Ampilokhiyev. Estimation of icebreaking properties of hovercraft. Proceedings of Marine Central Research Institute, 1979, Issue 243, pp.105-110.
138. S.P. Smigelsky. American works by program of developing arctic hovercraft. Sudostroyeniye za Rubezhom, 1978, No. 3 (135), pp-3-18.
139. V. Smirnov. Radiation channels and their use. Morskoy Flot, 1967, No. 1, pp.24-25.
140. Snow (and even ice) is blown off aerodrome by hot gas. Izobretatel i Ratsionalizator, 1970, No. 12.
141. A.T. Lupa, L.V. Ivanov, E.S. Vinogradov, L.A. Kapustin. Modern means for destruction and prevention of ice cover. Review Information of Central Bureau of Scientific-Technical Information, Sudoremont Series. Moscow, 1974.
142. A.V. Panyushkin, N.V. Sergachev, Yu.T. Sukhov, S.M. Aleinikov. Condition and development of anti-icing means. Proceedings of Coordination Conferences on Hydraulic Engineering, 1976, Issue 111, pp.3-9.
143. F.A. Spetsov. Weakening ice in spring by blackening its surface. Proceedings of Coordination Conferences on Hydraulic Engineering, 1965, Issue 17, pp.225-235.
144. V.M. Tavrizov. Use of ice cutting machine for construction of underwater crossings. Stroitelstvo Truboprovodov, 1964, No. 9, pp.31-33.
145. V.M. Tavrizov. Mechanized ice drill. Avtomobilniye Dorogy, 1966, No. 12, p.20.
146. V.M. Tavrizov. Ice drill with auger. Rybovodstovo i Rybolovstvo, 1971, No. 6, p.23.
147. S.S. Torban. Mechanization of River Seine Fishing. Moscow, Pishchepromizdat Publishers, 1955.
148. S.S. Torban. Mechanization of Fishing in Internal Water Bodies. Moscow, Pishchevaya Promyshlennost Publishers, 1969.
149. S.S. Torban, V.N. Danilchenko. Some parameters of ice drilling

process. Proceedings of All-Union Research Institute of Marine Fishing and Oceanography, 1959, Vol. 39, pp.91-98.
150. S.S. Orban, S.I. Polulyak. Tractor ice drill ILB. Rybnoye Khozaistvo, 1960, No. 12, pp.39-45.
151. S.S. Torban, S.O. Polulyak. Comparative testing of ice drilling units, type ILB. Rybnoye Khozaistvo, 1962, No. 12, pp.49-56.
152. I.D. Tkachevsky, A.A. Grinchik, I.V. Budanov, V.V. Gusyev. Removal of winter ice at filling flooded embankments at Baikal-Amur railway. Transportnoye Stroitelstvo, 1978, No. 9, pp.3-5.
153. Yu.V. Treushnikov. Systematization and Analysis of Means for Fighting Ice in the USSR and Abroad on Internal Waterways under Conditions of Prolonged Naviation. Gorky, 1968.
154. E.S. Troshkina. Artificial intensification of ice and snow melting. Engineering Glaciology. Moscow State University, 1971, Chapter VI, pp.125-131.
155. E.A. Trubina. Investigation of drilling ice by screw drills. Proceedings of Gorky Polytechnical Institute, 1970, Vol. 26, Issue 1, pp.40-44.
156. O.K. Trunov. Icing of Aeroplanes and Means for its Prevention. Moscow, Mashinostroyeniye Publishers, 1965.
157. G.P. Utkin, V.I. Kuklin. Method of removing ice from dredge pits. Kolyma, 1973, No. 2, pp.6-8.
158. L.I. Faiko. New method of artificial reduction of mechanical strength and heat conductivity of ice cover on water bodies. Materials of Glaciological Investigations. Discussion, 1973, Issue 21, pp.151-163.
159. N. Fedotenkov. New ice drill. Rybovodstvo i Rybolovstvo, 1971, No. 4, p.40.
160. V.I. Fedotov. Drilling tools used by foreign ice researchers. Proceedings of Arctic and Antarctic Research Institute, 1964, Vol. 257, pp.153-158.
161. V. Fokeyev. Ice destruction for navigational purposes. Rechnoy Transport, 1969, No. 4, pp.45-46.
162. D.E. Kheisin, V.A. Likhomanov. Experimental determination of specific energy of mechanical disintegration of ice at an impact. Problems of Arctic and Antarctic, 1973, Issue 41, pp.55-61.
163. D.E. Kheisin, N.V. Cherepanov. Change of ice structure at zone of solid body impact on ice cover. Problems of Arctic and Antarctic, 1970, Issue 34, pp.79-84.
164. N.G. Khrapaty. Methods of Determining Strength Characteristics of Ice. Vladivostok, Far East Polytechnical Institute, 1979.
165. N.G. Khrapaty. Calculation of depth of solid body penetration into an ice cover at an impact. Rational Utilization of Natural Resources and Environmental Protection. Leningrad, 1978, pp.124-126.
166. N.G. Khrapaty, V.G. Tsuprik. On an impact of a solid body on ice. Proceedings of Far East Polytechnical Institute, 1974, Issue 60, pp.105-108.
167. N.G. Khrapaty, V.G. Tsuprik. Flexural vibrations of ice cover caused by impact pulse. Proceedings of Far East Polytechnical Institute, 1975, Issue 109, pp.12-26.
168. N.G. Khrapaty, V.G. Tsuprik. Experimental investigation of an

impact of a solid body on ice. Proceedings of Coordination Conference on Hydraulic Engineering, 1976, Issue 111, pp.166-169.
169. N.G. Krapaty, V.G. Tsuprik. Experimental investigation of flexural vibrations of an ice cover at an impact. Hydraulic Engineering and Hydraulics, 1977, Issue 2, pp.74-77.
170. V.L. Tsurikov. On ice picks and ice drills. Rybolov-Sportsman, 1958, No. 10, pp.62-64.
171. E.N. Tsykin. Method of ice destruction by 'large chipping' and its place in system of anti-jamming measures. Materials of Glaciological Investigations, 1970, Issue 17, pp.309-316.
172. E.N. Tsykin. Weakening ice cover by furrowing. Proceedings of Coordination Conferences on Hydraulic Engineering, 1970, Issue 56, pp.70-73.
173. E.N. Tsykin. Use of ice ploughs to prevent ice jams. Transactions of USSR Academy of Sciences, Geography Series 1970, No. 3, pp.61-66.
174. E.N. Tsykin. Active effect on ice cover under conditions of shallow waters. Proceedings of Coordination Conferences on Hydraulic Engineering. Fighting Ice Troubles when Operating Hydrotechnical Structures, 1973, Vol. 81 (D), pp.151-154.
175. E.N. Tsykin, G.A. Tsykina. Results of laboratory experiments on enhancing artificially ice melting by blackening. Materials of Glaciological Investigations, 1968, Issue 14, pp.167-179.
176. G.A. Tsykina. Experiments on enhancing melting of ice cover under unfavourable weather conditions. Materials of Glaciological Investigations, 1970, Issue 16, pp.191-195.
177. G.A. Tsykina. Artificially enhancing melting of ice cover on water bodies. Materials of Glaciological Investigations, 1971, Issue 18, pp.68-72.
178. N.V. Cherepanov. New circular ice drill. Problems of Arctic and Antarctic, 1969, Vol. 32, pp.122-125.
179. E.I. Chestnov. Use of hovercraft to break up ice. Advanced Experience and New Technology, 1979, Issue 2 (62), pp.69-73.
180. I.N. Shatalina, F.A. Spetsov. Experience of using chemicals to weaken ice strength. Proceedings of Leningrad Institute of Water Transport, 1963, Issue 26, pp.59-65.
181. N.K. Shafranov, A.G. Kuznetsov, Yu.Yu. Glukhov. Specific features of rock destruction by the beam of a continuous-action laser. Physico-technical Problems of Developing Useful Minerals, 1978, No. 1, pp.47-51.
182. Z.I. Shvaistein. Ice cutting by continuous high-pressure jets. Proceedings of Arctic and Antarctic research Institute, 1971, Vol. 300, pp.168-176.
183. Z.I. Shvaistein. Ice destruction by high-pressure pulse jets. Proceedings of Arctic and Antarctic Research Institute, 1972, Vol. 331, pp.203-211.
184. Z.I. Shvaistein, R.K. Englin. Ice destruction by vibration. Proceedings of Arctic and Antarctic Research Institute, 1964, Vol. 267, pp.89-99.
185. A.I. Shkoda. Some prerequisites to estimation of operation effectiveness of ice cutting machines with end and side milling cutters. Proceedings of Gorky Polytechnical Institute, 1975, Vol.

31, Issue 8, pp.5-10.
186. A.S. Shchetinnikov. On problems of expediency of enhancing artificially the melting of glaciers in Syrdarya River basin. Proceedings of Central Asia Research Institute, 1977, Issue 13, pp.13-15.
187. G.N. Yakovlev. Investigations on problem of ice destruction. Proceedings of Arctic and Antarctic Research Institute, 1964, Vol. 267, pp.54-63.
188. G.N. Yakovlev. Ice destruction by reactive gas jet. Proceedings of Arctic and Antarctic Research Institute, 1964, Vol. 267, pp.54-63.
189. C.W. Young. Parametric investigation of under ice hydro-acoustical buoy. All-Union Centre for Translations of Scientific-Technical Documentation, 1977, No. 99315.
190. R. Yarov. Ice against ice. Izobretatel i Ratsionalizator, 1974, No. 10, pp.16-69.
191. Air bubbling. Technical Memorandum of the Association for Commercial Soil and Snow Machanics National Research Council, Ottawa, 1961, No. 70.
192. Air cushion ice breaking study. Canadian Shipbuilding and Marine Engineering, 1978, Vol. 49, No. 6, p.42.
193. Air cushion platform may improve icebreaker performance. Marine Engineering Log, 1975, Vol. 80, No. 9, p.5.
194. B. Arnason, H. Björnsson, P. Theodorson. Mechanical drill for deep coring in temperature ice. Journal of Glaciology, 1974, Vol. 13, No. 67, p.133-139.
195. K.C. Arnold. An investigation into methods of accelerating the melting of ice and snow by artificial dusting. Geology of Arctic, 1961, Vol. 2, pp.989-1013.
196. Bell's surface skimmers performing a wide variety of tasks. Marine Log, 1975, Vol. 80, p.33.
197. M. Broshu. Les vehicles a coussin d'air des brise-glace novelle vague. Science Dimension, 1977, Vol. 9, No. 6, pp.17-19.
198. M. Broshu. Possible use of light absorptive material for seasonal deicing off Alaska North Slope. Sea Technology, 1977, Vol. 18, No. 2, pp.15-17.
199. J. Browning. Flame drilling through the Ross Ice Shelf. North Engineering, 1978, Vol. 10, pp.4-8.
200. J.L. Coburg, N.A. Elrich. Advanced icebreaking concepts. Naval Engineering Journal, 1973, Vol. 85, No. 24, pp.11-19.
201. A.R. Clark, J.C. Moulder, R.P. Reed, Ability of a CO_2 laser to assist icebreakers. Applied Optics, 1975, Vol. 12, No. 6, pp.1103-1104.
202. Core obtained through Ross Ice Shelf. Antarctic Journal (U.S.), 1979, Vol. 14, No. 1, p.1.
203. L. Corsby, Ice engineering research at CRREL. Arctic Bulletin, 1977, Vol. 2, No. 10, pp.177-181.
204. I. Dalrymple. The Great Lakes: Canada's fourth ocean. Canadian Shipbuilding and Marine Engineering, 1977, Vol. 48, No. 8, pp.12-19.
205. Drillers put three holes through Ross Ice Shelf. Antarctic Journal (U.S.), 1979, Vol. 14, No. 1, pp.4-5.
206. M. Eldrup. Vacancy migration and void formation in irradiated ice.

REFERENCES

Journal of Chemical Physics, 1976, No. 12, pp.5283-5290.
207. M. Eldrup, O. Mogensen. Vacancies in HF-doped and irradiated ice by positron annihilation techniques. Journal of Glaciology, 1978, Vol. 21, No. 85, p.1011-1013.
208. R.I. Evans, N. Untersteiner. Thermal cracks in floating ice sheets. Journal of Geophysical Research, 1971, Vol. 76, pp.694-703.
209. R.I. Evans. Cracks in perennial sea ice due to thermally induced stress. Journal of Geophysical Research, 1971, Vol. 76, pp.8153-8155.
210. Fighting ice. Ocean Industry, 1978, Vol. 13, No. 10, pp.144-146.
211. First hovercraft tests on ice. Antarctic, 1977, Vol. 8, No. 1, p.8.
212. H.S. Fowler. The air cushion vehicle: a possible answer to some Arctic transport problems. Polar Record, 1976, Vol. 18, No. 144, pp.251-258.
213. F. Gillet, D. Donnon, C. Ricon. A new electrothermal drill for coring ice. Ice-core Drilling. Proceedings of Symposium, University of Nebraska, 1974, pp.19-27.
214. R.R. Gilpin. Hydraulic cutting of ice and frozen gravel. Summary of Current Research of Snow and Ice in Canada, 1976, p.12.
215. J. Golecki, C. Jaccard. Radiation damage in ice at low temperature studied by proton channelling. Journal of Glaciology, 1978, Vol. 21, No. 85, pp.247-258.
216. S. Hodge. A new version of a steam-operated ice drill. Journal of Glaciology, 1971, Vol. 10, No. 60, pp.387-393.
217. R.L. Hooke. University of Minnesota ice drill. Ice-core Drilling. Proceedings of Symposium, University of Nebraska, 1974, pp.47-57.
218. F. Howorka. A steam-operated ice drill for the installation of ablation stokes on glaciers. Journal of Glaciology, 1965, Vol. 5, No. 41, pp.749-750.
219. Iceberg demolition experiments 1959. U.S. Treasury Department Coast Guard Bulletin, 1960, No. 45, p.2.
220. Icebreaking ACV platform stirs oversea interest. Canadian Shipbuilding and Marine Engineering, 1977, Vol. 48, No. 5, p.37.
221. Icebreaking. Bubble-blowing catches on. Canadian Shipbuilding and Marine Engineering, 1975, Vol. 46, No. 9, p.5.
222. Icebreaking hovercraft. Canadian Shipbuilding and Marine Engineering, 1973, Vol. 44, No. 9, p.5.
223. Ice melting machine. All Hands, 1961, Vol. 3265, No. 534.
224. C. Irwin. Analysis of stress and strains near the end of a crack traversing a plate. Transactions of the American Society of Mechanical Engineers, Journal of Applied Mechanics, 1957, No. 9, pp.361-364.
225. J. Lane, S. Marshall, Deicing using lasers. Meteorological and Geoastrophysical Abstracts, 1977, No. 8, p.1483.
226. J.F. Lea. A concise method for analyzing the ice-melting performance of a heated discos. Journal of Glaciology, 1976, Vol. 17, No. 76, pp.355-358.
227. F.I. Legerer. Mechanics of icebreaking. Summary of Current Research on Snow and Ice in Canada, Ottawa, 1976.
228. H. Liu, L. Loop. Fracture toughness of fresh-water ice. CRREL Draft Report, 1973.
229. W.K. Lyon. Experiments in the use of explosives in sea ice. Polar

Record, 1960, Vol. 10, No. 66, p.237-247.
230. M. Mellor. Cutting ice with continuous jets. Proceedings of International Symposium on Jet Cutting Technology, Cambridge, 1974. Cranfield, 1974.
231. M. Mellor, P. Sellman. General considerations for drill system design. Ice-core Drilling. Proceedings of Symposium, University of Nebraska, 1974, pp.77-111.
232. M. Mellor, A. Kovacs, J. Hnatiuk. Destruction of ice islands with explosives. Proceedings of the 4th International Conference on Port and Ocean Engineering, Newfoundland, 1977, Vol. 2. Newfoundland, 1978, pp.735-765.
233. T. Messer. Seaway icebreaker role for ACV. Canadian Shipbuilding and Marine Engineering, 1976, Vol. 47, No. 4, p.21.
234. O. Mogensen, M. Eldrup. Vacancies in pure ice studies by positron annihilation techniques. Journal of Glaciology, 1978, Vol. 21, No. 85, pp.85-98.
235. M. Mohaghegh. Fracture of sea ice sheets. 6th Annual Offshore Technology Conference, Houston, 1974, Vol. 2. Dallas, 1974, pp.121-132.
236. E.R. Muller. Icebreaking with an air cushion vehicle. SIAM Review, 1979, Vol. 21, No. 1, pp.129-135.
237. M.V. Arctic the first of how many? Canadian Shipbuilding and Marine Engineering, 1978, Vol. 49, No. 10, pp.25-29.
238. R. Naruse, Y. Susuku. A steam-operated drill used by the 14th Japanese Antarctic Research Expedition (1972-1974). Antarctic Research, 1975, No. 53, pp.53-56.
239. I. O'Nayl. Icebreaking helped by ACV technology. Canadian Shipbuilding and Marine Engineering, 1976, Vol. 47, No. 6. pp.29-30.
240. H.C. Parsons, R.M. Hopkins. A chemical method for ice destruction. Proceedings of the International Conference on Port and Ocean Engineering, Newfoundland, 1977. Vol. 2. Newfoundland, 1978, pp.799-810.
241. C.R. Payton. Some mechanical properties of sea ice. Ice and Snow. Properties, Processes and Applications. Cambridge, 1963.
242. W.S. Paterson. Thermal core drilling in ice caps in Arctic Canada. Ice-core Drilling. Proceedings of Symposium, University of Nebraska, 1974, pp.113-116.
243. H.J. Rand. The USA CRREL Shallow Drill. Ice-core Drilling. Proceedings of Symposium, University of Nebraska, 1974, pp.133-137.
244. Regelation of ice. Nature (Physical Sciences), 1973, Vol. 242, No. 114, p.1.
245. The submerged ice-cracking machine. Canadian Shipbuilding and Marine Engineering, 1972, Vol. 43, No. 8, p.16.
246. 687-ft icebreaking bulk carrier to be built in Canada. Marine Engineering Log, 1975, Vol. 80, p.10.
247. R.A. Smith. The application of fracture mechanics to the problem of crevice penetration. Journal of Glaciology, 1976, Vol. 17, pp.223-228.
248. P.L. Taylor. Solid nose and coring thermal drill for temperature ice. Ice-core Drilling. Proceedings of Symposium, University of **Nebras**ka, 1974, pp.97-99.

REFERENCES

249. Technical notes: vibrating ice. Marine Engineering Log, 1972, Vol. 77, No. 5, p.5.
250. D.R. Topham. The deflection of an ice sheet by a submerged gas source. Transactions of the American Society of Mechanical Engineers, 1977, series E, Vol. 44, No. 2, pp.279-284.
251. Trawler deicing equipment. Shipbuilding and Shipping Record, 1969, Vol. 113, No. 12, p.401.
252. R.G. Wade. Icebreaking. Air cushion technology promises a cheap way. Canadian Shipbuilding and Marine Engineering, 1974, Vol. 45, No. 11, pp.19-20.
253. R.G. Wade. Air cushion vehicle icebreaking. Summary of Current Research on Snow and Ice in Canada. 1976, Technical Memorandum No. 118, Ottawa, 1976.
254. R.G. Wade. Air cushion technology. Ice, 1978, No. 56-57.
255. W.G. Wood. Discussion on paper: Impact of spheres on ice. Journal of the Engineering Mechanics Division of the American Society of Civil Engineers, 1970, Vol. 97, No. 4, p.1293-1296.
256. Y. Yin-Choo, O. Fout, B. Lynnwood. Impact of spheres on ice. Journal of the Engineering Mechanics Division of the American Society of Civil Engineers, 1970, Vol. 96, No. 5, p-641-652.
257. C.W. Young. Penetration of sea ice by air-dropped projectiles. Ocean'74 IEEE International Conference on Engineering Ocean Environment Research, Halifax, 1974. New York, 1974, 1/80-1/95.

SUBJECT INDEX

Accelerated breakup of ice sheet   124
Adhesion of ice   179
Adsorption-surface effect   179, 182
Air-cushioned transport (ACT)   19
Anti-slush reagent   179

Blackening   124-127
Bubbling   128, 131-134

Chemical destruction   177
Chipping   23, 47
Combined destruction   191
Contact stress   5
Continuous-action laser   186-187
Crack propagation speed   26
Cutting means   26

Discharging warm water   141
Drilling   51

Effectiveness of top melting of blackened ice   126
Effectiveness of bubbling   133
Ejector device   138
Electic hydraulic technique   185-186
Electric pulse anti-icing system (EPAS)   182, 185
Electric puncture of ice   185
Electric thermodrilling of ice   161
Electric thermomechanical method   194
Electromagnetic field   187-188
Electrophysical destruction   184
Eutectic solution   177
Explosive methods   92

Flow generator   127, 136-140
Fragile destruction   24

Gas-thermal means   153

High-frequency melting   188
High-temperature gas   153
Hydrodynamic plants   136
Hydraulic erosive effect   136

Icebreaker oil tanker   192
Icebreaking system   8
Ice-cutting units   29
Ice drill   59
Ice-milling machines   34
Initiation of cracking   5

Kinetic concept   1

Laser ray beam gun   187
Laser technique   186
Layered structure in ice cover   142

Mechanical concept   1
Mechanical destruction of ice   1
Mechanical disintegration   3
Melting of ice in eutectic solution   177
Milling   23, 34
Modulus of normal elasticity   3

Natural reserves of heat in water   128, 140

Organo-silicate coating   179-180

Perforating   3
Pneumatic units   128
Prevention of ice formation   145, 171

Radioactive irradiation   189
Reactive burner   153
Rebinder effect   179, 182
Reducing mechanical strength of ice cover   142
Reflectivity of ice   124
Rotating scraper   50

Sinking of floes   192, 194
Solar radiation method   124
Steam-water-air methods   147
Strength of ice   1
Strength of irradiated ice specimens   189

Thermodrill   158
Trepanning ice drill   67

Underice acoustic buoy   5
Underwater gas exhaust   15

Vibration devices   11

Water body aeration   128
Water stream   89

RAYMOND H. FOGLER LIBRARY
DATE DUE

BOOKS ARE SUBJECT TO RECALL AFTER TWO WEEKS

~~NOV 2 5 1988~~